D1372697

BLOND'S TEACHERS' HANDBOOKS

Geography in the Field

Edited by
K. S. Wheeler

Senior Lecturer in Geography
City of Leicester College of Education

BLOND EDUCATIONAL

Also in this series: History Handbook ed. John Fines:
History in the Field ed. Tom Corfe: Religious Studies
ed. Joan Tooke. See also Leicestershire Landscapes ed.
K. S. Wheeler, case studies in local geography for the
C.S.E. pupil.

First published in Great Britain 1970
By Blond Educational,
56 Doughty Street,
London W.C.1.

219 51740 1

Printed in Great Britain by
Clarke, Doble & Brendon, Ltd.,
Plymouth

Contents

List of Illustrations

Introduction

This book has evolved from the work of a group of Surrey teachers, who under the auspices of the County Inspectorate, formed in 1958 the Surrey Fieldwork Society. The original Handbook, issued by Surrey County Council, met with a ready demand. It was therefore re-edited and amplified, and published by Blond Educational in order to serve a wider public. The first edition was welcomed by teachers in all parts of Britain, because it gave guidance on the methodology of fieldwork which is applicable nationally. It is now published again in a second edition.

The aim of the second edition is to extend the methodology of fieldwork in its practical application to the teaching situation. At the same time it retains most of those aspects of fieldwork dealt with in the earlier edition, which have proved popular among school teachers seeking assistance in preparing pupils for outdoor study of the environment.

The desire to provide a wider coverage has led the editor to bring in contributors from outside the Surrey Fieldwork Society. In particular, he is grateful to the Scottish Field Studies Association for enabling fieldwork to be included which illustrates two geographical excursions that can be made from their field study centre at Blairgowrie, Perthshire. This has meant that the fieldwork techniques explained in the book are illustrated not only with examples from Wales and many parts of England but also from Scotland.

The editor wishes to thank Miss A. E. Adams, Secretary, Surrey Schools Council, who inspired and constantly supported the work of the Surrey Fieldwork Society; Mr R. H. Sanders who prepared the diagrams for all the editions of the Handbook; the Managing Director of Invicta Plastics for permission to include the farm study; *Teachers' World* for agreeing to the publications of the two articles in the appendix; also Mr F. K. Balfour for permission to publish the account of the Highland Estate. In addition, acknowledgement is due to the numerous teachers, students, and pupils who have given much direct or indirect help. Among these are Mr John Emery, Lecturer in Geography, Sydney Teachers' College, Australia; Mr Bryan Waites, Senior Lecturer in Geography, City of Leicester College of Education; Messrs M. Baxter, A. Burrows, P. Coatham and M. Walton, and Misses S. Brown, M. Guest, G. Pelland and S. Wolowitz. Finally, the editor would like to thank Mr James Kenyon of Blond Educational for his constant help and patience.

It only remains to be said that any errors or omissions are the responsibility of the editor.

K. S. Wheeler

1

Geography Teaching
and Fieldwork

Geography teachers were among the first educationists to develop their teaching as an exploration of the reality around them. The work of teachers such as Fairgrieve brought the actual world into the classroom by the description of communities and places as they actually exist, and from this was born the method of using sample studies. As early as 1913, A. J. Herbertson wrote: 'If geography did nothing but teach a child to see, know, and love his own district it would be an inestimably valuable element in education. But it can do more. It can lead him to the conception of varieties of the home type district and from that to appreciate in part some of the outstanding characteristics of quite different types of district'.

The development of fieldwork techniques also arose out of the geographer's ready appreciation that the best experience of geographical reality could be had by making investigations into the environment outside the classroom. Indeed, by the early part of the century, field excursions were a familiar feature in several schools, but not until after the Second World War did fieldwork become a more generally accepted practice. Even today, however, there are still many teachers who are unable to use fieldwork in their teaching either because of their lack of experience or because of organisational difficulties arising in the school.

It is salutary that contemporary developments in education have confirmed the basic correctness of this approach, despite the fact that many geography teachers have not always been able to put it into practice. Teaching through discovery, proceeding from the known to the unknown, learning through experience – these are the canons of present day educational practice. Yet the irony is that while geography teaching at its best has advanced far in this direction, the great mass of geography has been taught purely as a means of giving children a compendium of factual information about the world. Many geographers have clung to the descriptive regional method for putting over their subject in the classroom, despite the more enlightened approaches that can be used. This is seen for instance in the number of geography syllabuses in secondary schools which continue to attempt a world coverage. There is often an inordinate rush to hurry children on from one continent to another, frequently squeezing out field-work because of lack of time, and filling the pupils' minds with a simplified corpus of factual knowledge. As Dr G. M. Hickman pointed out in *Developments in Geography* (Papers from a Conference for Secondary School Teachers: Moray House College of Education): 'Until we agree to teach less in order to teach more effectively, I do not believe that we shall allow geography to make its full contribution'.

It is the factual, descriptive content of geography, usually couched in a regional formalisation, which has laid a retarding hand on the development of the discovery method. As Professor N. V. Scarfe once declared: 'Geography should be a light in the mind, not a load on the memory'. But how can this be achieved? Firstly, by emphasising geography as a practical subject, the teaching of which can be best carried out by activity methods. This means making fieldwork a fully integrated part of the syllabus, and equipping and using the geography room as a laboratory. Secondly, by emphasising that it is the 'geographical point of view' which it is important for the pupil to acquire rather than the factual content of the subject. This point of view is one that attempts to see the relationships existing between natural and human phenomena in a spatial context. It is geography as a bridging subject, cross-linking the sciences and humanities, that gives it its important role to play in the education of modern youth. Indeed, this is the great intellectual birthright inherited from the thinking of the founders of modern geography, Humboldt and Ritter, which is embodied in the concept enunciated by them *zusamenhängen* or 'the hanging together of things'. In a modern technological age which relies too heavily on analytical ways of thinking, geography provides a corrective by viewing the land environment as a unity.

This fundamental characteristic of geography is in accord with the contemporary development in education which rejects early specialisation in discrete disciplines as being contrary to the way children best learn. As A. N. Whitehead wrote in his book *'The Aims of Education'*, 'the craving for expansion, for activity, inherent in youth is disgusted by the dry imposition of disciplined knowledge'. The whole spirit and purpose of geography can provide a discipline which is neither dry nor imposed, but vital and significant, provided the educational methods used are a true reflection of its declared aims. In order to do this the geography teacher must have a profound appreciation of his subject and a deep understanding of the processes of learning. Indeed, he must develop his educational role as an 'entrepreneur', placing children in learning situations which allow them to inter-relate enquiries across subject barriers in their efforts to achieve a better understanding and feeling for environment than has been the case until now. Both in the geography room and in the field, the teaching of geography must be directed more to the solution of problems than the acquisition of fact. This is why, for instance, fieldwork should not be conceived of as a guided coach tour, or purely as an out-of-door lecture. Professor R. E. Dickinson delineated the kind of problems suitable for geographical investigation in this way: 'The problems of geography are real problems. The terms areal, spatial and geographic are synonymous. The use of the term geographical as synonymous with physical is a hangover from the nineteenth century viewpoint, and, in our view, quite categorically, should be dropped. A geographic fact is any areal fact, physical or cultural, that is relevant to a particular problem.' (*Some Problems of Human Geography*, Leeds University Press, 1960.)

It was the late Professor S. W. Wooldridge who charted with insight the relationship between geography and education. His thoughts on this are contained in his book, *The Geographer as Scientist*. It was he who gave the most succinct definition of geographical fieldwork as: 'the art of seeing and using accessible local ground as a laboratory for our teaching'. He declared further: 'The geographer, by temperament and outlook catholic, and no believer in subject autonomy, will not be disinclined to co-operate with historians or to join the family of social studies'. Then in a famous

2

passage he cautions: 'What I chiefly fear if the narrower humanities gain control, is that the physical elements of geography will be unduly subordinated'. In other words, he was fearful lest 'the ge- be taken out of geography and the baby thrown away with the bath water'. In fact, too, any move away from the kind of amalgamation he described could also lead to a state of affairs where teaching about maps, which surely is a special function of school geography, might well be reduced, and the world outlook attempted by geography limited to a curriculum far too parochial in its scope.

As it is, geography as a discipline is becoming increasingly scientific. The application of quantitative techniques, for instance, developed in Universities in recent years, is beginning to influence the schools. In many ways this can be welcomed as emphasising the problem-solving character of geography; but at the same time there is a threat that geography may become a 'faceless' subject, in which the human and aesthetic elements of the environment are ignored. The paradox is that this trend is occurring at the same time as there is a move away from specialisation in the schools and a desire to see geography absorbed with history and sociology into a unified course. Such a situation threatens to place considerable strain on the teacher because the increasing pressure from the University to 'quantify' clashes with the desire to break away from subject teaching at the grass roots of the educational system.

It is possible that the swing towards the scientific has gone too far; certainly physical geography may hold too dominant a position, distorting the total geographical point of view. (Although there is also a lot to be said for G. E. Hutchings' opinion that, 'no prospectus of geographical field teaching is complete without the provision for the study of plant and animal ecology'. This is an aspect sadly neglected). Rather may it not be more the case that geography teaching, both in the field and in the geography room needs to become more relevant to the environmental problems that human beings are experiencing in the contemporary world? As Mrs M. Long writes in the introduction to the *Handbook for Geography Teachers* (Methuen, 1964): 'It is clear that geography still studies places; the study of people in them appears diminished in emphasis. It needed Wooldridge in 1949 to insist on the 'ge' in geography. Perhaps our need now is for a like champion to put geography again among the humanities'.

It is not only a question of putting geography back among the humanities, but also of re-asserting the character of geography as the study of the whole spatial environment: physical, biological, human. If this is so, the responsibility falls upon the geography teacher to examine more closely the way children actually learn. Professor Wooldridge averred: 'The greatest obstacle to the better teaching of geography is ignorance of geography i.e. the matter of the subject, not, be it noted, of child psychology'. The first part of his statement is undoubtedly true, as is emphasised here, but was the Professor right in implying that we should turn our backs on the findings of child psychology? The advancement of modern methods of teaching has come about because of a greater awareness of the way children learn. Indeed this is a justification for employing fieldwork as a teaching method, not merely because it is good geography, but also because out of it can come the self-activation of the pupil who learns because he finds the learning pleasurable and interesting.

There is a great need to assess the development of geography teaching in relation to

3

defined curriculum objectives so that the many-sided method of teaching the subject may be adapted according to the requirements of children of diverse ability and level. The emergence of environmental studies in primary school teaching illustrates a move in this direction. At this stage the children are holistic in their attitude to the world around them and it is possible to introduce a wide range of environmental phenomena into a single area to stimulate learning interest. In the middle school however learning may be pursued at greater depth without congealing the young mind into a rigid subject mould. This may well be avoided by using the developing techniques of team teaching, and possibly the centering of projects around more problem-solving enquiries than hitherto. The fieldwork approach should reflect this and the teacher must not be afraid of stepping outside the narrowly defined limits of what is and what is not geographical fieldwork. Similarly with the sixth form: the pupil who is not intending to study geography at university may require a course containing many of the elements of geography, but one in particular which will develop an ability to look beyond the horizons of his own preferred subject. Such a course might be one dealing with landscape appreciation, ranging from the scientific to the aesthetic. In short, we should not be afraid to remodel our geography teaching in order to accommodate the diverse needs of schoolchildren at different levels in their school career.

This book is mainly concerned with fieldwork for the older primary pupil, and the secondary child up to the age of fifteen. In addition it has suggestions for fieldwork investigations for older pupils who are following a more general course. It is essentially a book by teachers for teachers. It has attempted to provide methods of approach to a wider diversity of fieldwork than before. It also indicates links between work in the field and the use of 'documentary' evidence in the library or the geography room. Above all it is offered in the hope that it will continue to help teachers and school children to enjoy and develop a geographical point of view in understanding the environment.

2
Fieldwork with Schoolchildren

The aim of fieldwork with children is to place them in a practical learning situation which involves making observations of a geographical nature. In doing this the children should be encouraged to seek correlations in the patterns of phenomena they observe in the field, thereby increasing their understanding of the inter-relationship in the physical environment.

In practice, fieldwork proceeds in three connected stages. These are:

1 The preparation for fieldwork
2 The fieldwork excursion
3 The synthesis of the observations made.

1   The Preparation for Fieldwork

The key to the success of school fieldwork is long-term planning and good organisation. Those who sally forth unprepared into the field, with classes similarly unprepared, court disaster.

First, the teacher must become familiar with the area in which his school is situated. Upon joining a new school he should acquire a personal set of local maps and plans for study and reference. Maps are meant to be annotated, improved, added to, and generally written on, preferably in indian ink. The teacher's private set of maps should soon become smothered in a mass of personal jottings as details observed in walks round the school and home area are recorded. Clipped to the maps, or filed with them, will be sketches, diagrams and lists of rock exposures, viewpoints, museums, etc.; picture postcards, photographs or pamphlets, admission charges and bus time-tables, and in fact, all snippets of information that might be of future use. He must spend time reading relevant books and documents from local museums and libraries and local government offices, etc. Fieldwork should be the teacher's interest in and out of school. He must do, and be seen to do, everything the children are expected to do, e.g., soiling hands in the testing and collecting of rock and soil samples.

Second, he should decide where fieldwork can be related to the school geography syllabus. Much will depend on the age of the children concerned and their level of attainment. Ideally, however, fieldwork should develop out of the sequence of lessons in the classroom. It should not be an activity which is arbitrarily imposed on the school work, but should exemplify preceding theoretical work or be a starting-point from which lessons are developed. It must be an experience which generates an appetite for more geography: fieldwork *is* fun.

Fieldwork can take a period, half a day, a whole day, or more. In simple cases it can be done without leaving the school premises. For example, a fieldwork programme for a secondary modern school in an urban area started in the first year with short

5

investigations introducing basic techniques, and ended in the fourth year with a larger investigation requiring several days to complete.

A Four-Year Programme
of Fieldwork

First Year

Orientation of map and map traverse	2 double periods
Viewfinder and urban diagram exercises from top window of school	1 double period
Farm study	1 day

Second Year

Simple stream study combined with simple traverse across local common	1 morning
Investigation of site of mills on local river	1 morning

Third Year

Study of a local parish to show siting and growth from an Anglo-Saxon settlement (in conjunction with history teacher)	1 morning
Survey of local trading estate. Group work	1 morning

Fourth Year

Traffic census and communications study of local borough. Group work	2 mornings
Traverse across borough to show origin, growth and development	1 morning
Traverse and transect across North Downs	1 day

The teacher having planned ahead by drawing up a fieldwork syllabus is now able to organise in preparation for the excursions. Consultations with the head teacher, requesting time during school hours for this work, are more likely to succeed if the place of fieldwork in the whole geography syllabus can be demonstrated in advance. Colleagues who as a result will stand to lose their specialist lessons with the classes involved are more likely to agree to the suggestion if they are approached well beforehand.

Finances: The problem of financing the more distant excursion is also eased by thinking well ahead. Most authorities give an allowance to schools for the purpose of school visits; money may also be available out of petty cash or school funds. An early claim to these possible sources of money has an obvious advantage. The children too can be encouraged to contribute to the fares on a saving basis over an appropriate period of time.

Equipment: This must be assembled before the fieldwork starts. The following list is a guide to the kinds of things necessary:

(a) Maps are essential – the 6″ and 2½″ scales being the most useful to have in quantity. It is extraordinary how few schools have enough maps. They are much more valuable and much cheaper than many textbooks.

(b) A spirit duplicator is most useful and senior pupils should be allowed to use it. This encourages them to produce results worthy of duplication for distribution to their fellows.

(c) A stiff-covered pocket-size field notebook, map case or piece of plywood or hardboard with bulldog clip, elastic bands, and plastic sheet or polythene bag for protection against rain, rubber eraser and HB pencils are needed for each pupil working in the field.

(d) Coloured crayons, duplicating carbons and paper, mapping and writing materials are all required for follow-up work in the classroom.

(e) Standard keys for land use surveys, geology and urban surveys should be available for reference. These will, of course, be simplified as required, and in any case children will first be encouraged to provide their own solutions. Only later will the value of standard practice be pointed out.

(f) A bottle of dilute hydrochloric acid with a dropper is the simplest soil-testing outfit. A bottle of soil indicator is also useful. This can be ordered at most chemists. More complicated soil-testing outfits (e.g., British Drug Houses set) are useful for advanced students. Long-handled trowels are an asset for collecting soil samples. Soil augers can be made by school metalwork departments, though a spade is more useful.

(g) Measuring chain and tape are essential for elementary surveying. Surveyors' poles can be made from suitably painted broom handles.

(h) A compass, though an accurate scientific instrument, can be purchased from stores dealing in government surplus stocks. Leaders of fieldwork expeditions in mountainous areas should always carry a compass. Silva compasses are quite adequate for school purposes.

(i) Abney levels can be obtained from the same source as compasses but any amateur can make a home-made clinometer to measure slopes – a simple protractor and bob line suffice. See the Appendix.

(j) Jam jars filled with paraffin are best for storing samples of moist clay. Polythene bags are useful for storage of soil samples.

(k) Any good hammer will be adequate for breaking off samples of chalk and wealden sandstones. Where harder rocks are likely to be encountered, it is safer to use a geological hammer.

(l) Rock samples can be collected and wrapped in newspaper or kept in polythene bags.

Duplicated handouts: These may consist of base maps, questionnaires, transect diagrams prepared for distribution to the pupils. Senior pupils should also be allowed to use the duplicator for this purpose. If possible, the teacher should run off at least twice the amount needed. This allows for children to make fair copies on a second sheet, or for a teaching set to be retained for the following year. A copy should be kept for the record with written comments on what improvements could be made next time.

Planning the route: The teacher should walk over the proposed route well beforehand. By having a first-hand experience of the area to be investigated the teacher can decide how best the fieldwork can be presented to the pupils. Important criteria are: can the children concerned appreciate the geography of the area? For example, a complex river study is not suitable for junior children but a local gulley or small stream might prove a worthwhile study. Is the walking distance within the competence of the children? How long will it take? (A fieldwork excursion usually takes longer with students than when alone.) What maps will be required?

Travelling: Where the fieldwork takes place some distance from the school it may be necessary to hire a coach or go by train. In the former case the order should be placed with the conveying firm after comparing quotations from elsewhere as far in advance as possible to ensure that a vehicle is available. In the latter case party tickets and circular route fares help to keep prices down.

Methods of investigation: Decide upon the best method by which your children will be able to work. Will this be a party led by the teacher? Can you send the children out in small groups on their own? Will they work in pairs or groups when in the area? Will they need to be put down at different starting-points?

Usually the ratio of pupils to teacher should not exceed twenty. If more need to go at one time, then student teachers, if available, are often keen to accompany a party; or a colleague who has historical or biological interests may be co-opted.

Detailed instructions: If the fieldwork is to take place away from school and to occupy a whole day or more, a duplicated sheet should be provided in advance, giving the itinerary and equipment and suggesting also suitable clothing, shoes and food to be brought. This is for the parents' as well as the pupils' information.

2   The Fieldwork Excursion

If ample preparation has been made and the children are clear as to the purpose of the investigation being made, and where it relates to their geography as a whole, then the fieldwork excursion should 'run itself'.

It is best to give out handouts and maps when they are required, and not at the outset of a journey. This prevents unnecessary loss or damage.

The teacher should avoid giving an open-air lecture when in the field. This defeats the main purpose of the exercise and causes restlessness among younger students. The teacher's task, when he is leading the party, is to activate interest, direct attention, follow up suggestions from pupils concerning observations they may make, answer questions, set the example. Emphasis should be on the children acting upon rather than listening to, the words of the teacher.

Note taking: An orderly method of taking notes in the field should be encouraged. Each child should number his notebook pages consecutively. The date and place of the fieldwork is recorded first as a heading. Thereafter it is useful if each note made is also numbered progressively throughout the day, and a six-figure map reference given alongside to show where the observation was made. If specimens are collected they should be labelled alphabetically. When photos are taken these should be recorded alongside the appropriate number of the observation.

3   The Synthesis

This is the final and important stage when the pupil, having experienced the geographical processes of observing, recording and interpreting, now attempts to draw all the threads together into a complete whole. In a very true sense, fieldwork does not become geography until the difficult write-up has been attempted. However, the synthesis may take many forms according to the age and ability of the children involved, and also the degree of complexity of the fieldwork itself.

It is important, at whatever level of attainment, that the collating together of the various investigations made should come as soon as possible after the day of the excursion while the experience is still fresh. Whenever possible, the pupils should be encouraged to write up their excursion in diary form the same evening.

With non-academic children, synthesis may simply take the form of a discussion based on the observations made, and supported by the making of a map from sketches done in the field. With all levels of school children the accurate completion of a transect diagram could be sufficient in itself. Nevertheless, the soundest method to aim for is the write-up, supported by maps, diagrams, specimens, photos, etc. and illustrative models. The 'z-book' method is simple but useful for the presentation of the write-up.

A z-book consists of a suitable thin board folded or joined to produce an extendable concertina shape of stiff enough paper or card on which the account of the fieldwork can be mounted and displayed. In assembling this the emphasis should be placed on putting the maps, diagrams and photographs in the order required and arranging round these the written material which explains the illustrations. When appropriate, small specimens such as dried plants or samples obtained from a factory can be attached to the pages of the z-book.

The advantage of this method is that it stimulates group work in assembling the finished account. The card provides a wider area than the notebook or loose-leaf pages, thus making display an easier matter, and the z-book is extendable by simply adding further sections. It can also be folded up and handled like a book (although opening differently) or stood up vertically for display purposes.

It is a stimulating and worthwhile experience for the children to put on a display of their work at some time in the year. In working towards a communal goal they will have the satisfaction of communicating the results of their investigations to a wider audience. Such a display could consist of the work actually completed by the children over the period of a year, or less. Parents can be invited to view the exhibition. It also has the great advantage of sparking off interest and added co-operation from parents and pupils alike. Similarly, when a survey has been made of a local area it can be a pleasant social occasion to invite people from the locality to inspect the work. Whether a display is put on or not, the children should always be given ample opportunity to read and examine the fieldwork results of their own fellows.

4 Assessing Fieldwork

In the case of a completed body of fieldwork which has been 'written-up' by the older pupil there is a considerable difficulty for the teacher in evaluating the standard which the student has achieved. It is necessary to consider the objectives of fieldwork, and to relate these to an assessment sheet which the teacher can complete and insert into the student's work. Here is an example of such an assessment sheet used for evaluating the results of a week's fieldwork. It should be noted that by no means does all fieldwork need to be written-up. It would take far too long, and make the whole exercise unduly laborious. It is better if the student can choose those aspects of the work in which he is interested most, and for the rest show the record of his work as it appears in rough in his notebook.

Assessment Sheet
A Fieldwork Excursion to the Gower Peninsula

Name of student...

Objective   *Grade and comment*

1 *Observation*
    (a) Accuracy
    (b) Originality
    (c) Interpretation
    (d) Correlation

2 *Presentation*
    (a) Written
    (b) Mapwork
    (c) Diagrams
    (d) Photographs
    (e) Field sketches
    (f) Specimens

3

Fieldwork and Maps

Maps are the primary aid to fieldwork. Indeed, the need to become familiar with map using and map making is a major reason for carrying out fieldwork. All children should leave school not only knowing how to use maps but also how to enjoy them. Yet it is often the case that schools do not possess an adequate set of topographical maps covering their home area.

In fieldwork investigations a knowledge of simple map making is required for recording observations made in the field, and a practical understanding of map using is wanted if our pupils are to improve their appreciation of the landscape. Properly used, maps go a long way to aiding the comprehensive view of the landscape, helping to see 'over the hill', as it were, and to discern spatial patterns within the environment, as well as helping us to find our way about.

1 Maps that can be used

(a) *Historical:* These are the maps kept by the local library or County Records Office. They provide documentary evidence for the character of the part landscape. They can be consulted to corroborate findings in the field or they may provide starting-points for investigation. For instance, present-day roads and boundaries often have their explanations in the past. Tithe maps and enclosure maps are particularly useful for finding out about the landscape in the late eighteenth century and early nineteenth century. The former may be preserved by the local church or Records Office but can also be consulted at: Tithe Redemption Commissioners Office, Finsbury Square, London.

Eighteenth-century topographical maps, such as Rocques maps, are also invaluable where they are available. Sometimes photographed copies can be obtained for handling in the classroom, or individual children can be sent on an assignment to study the original.

(b) *Ordnance Survey maps:* In the first place older editions of O.S. maps that are sometimes found lingering in school stockrooms should not be despised. They make excellent starting-points for investigation in land-use changes since the date they were published.

It is best to obtain teaching sets of the latest editions of the 25″, 6″, 2½″ and 1″ to 1 mile O.S. maps covering the school locality, buying them if necessary over a period of time. Where funds are limited the 2½″ maps are the most useful for teaching purposes with one or more copies of the local 1″ map to which they can be related.

These maps can be ordered from: Ordnance Survey, Romsey Rd., Maybush Southampton SO9 4DH. A discount of $33\frac{1}{3}$ per cent is allowed to educational institutions.

11

(c) *Geological maps:* At least one copy of the $1''$ map covering the school area is invaluable, along with the relevant Regional Geology Memoir. These can be had from The Director General, Ordnance Survey, or from the Geological Museum in Exhibition Road, South Kensington, London. The teaching of the geological time scale should be part of all classroom work in secondary school geography accompanied by use of the map to understand land forms and aid in the collection of rocks and specimens of fossils by the school and by individual pupils.

(d) *Other maps:* These consist of guide maps, road maps and the like. The former are particularly useful as a basis for urban fieldwork. Town plan maps illustrative of the local development plan are also useful.

(e) *Base maps:* The use of base maps is a fundamental technique of geographical fieldwork. The purpose of using a base map is: (a) for the teacher to simplify a more complicated map so that it can be duplicated for use in the field. (b) For the student to record results on after the teacher has prepared questions for him to investigate in the field.

In this way the student is free to carry out the fieldwork exercise independent of the teacher, and not as one of a party following the teacher around. The student should also be encouraged to make original observations, beyond those required of him by the base map, which he will also record. Finally, a fair copy of the base map can be prepared either as part of the write-up of the fieldwork, or as a large visual aid to be used for display purposes. Thus the base map has a similar value to the transect diagram (described in Chapter 8) for bringing together information gathered during fieldwork because its completion leads the student to correlate his observations in written, diagrammatic, and map form as is illustrated in the example given in Figure 1. This also shows that most of the questions given with this base map exercise and listed here were answered, and additional information added on the student's own initiative. The field sketch and photographs are recorded correctly, and would be submitted along with the completed work.

Base Map Exercise (to go with Figure 1)
FIELD QUESTIONNAIRE FOR DAY 2
Fall Bay to Rhossili Bay

High Tide: 10 a.m. and 10 p.m.

This guide to your day's fieldwork is not meant to cover all the possibilities for observation. Credit will be given for making further original enquiries.

i. Collect and identify examples of fossil-bearing rock.

ii. Look for old limekiln and quarry marked A and B on the base map. Explain their presence.

iii. Where possible examine sections of boulder clay deposits.

iv. Look for and record features of coastal erosion.

v. Mark direction of dip of rocks on the base map. To what extent are these consistent with the known local folding of the strata?

vi. Find and record example of possible fossil swallow hole revealed in the cliff section. What is your opinion of this feature?

vii. Find evidence for changes of sea level shown by the difference in the heights of caves above shore level.

viii. Find and record the location of the Patella raised beach. Collect specimens of the beach material. For what observational reasons do you assign this beach to a pre-glacial period?

ix. Observe and record the view at C looking towards Worm's Head. Try to account for this formation from observation and what you know of local folding.

x. Observe and explain the field patterns at D.

xi. Look for evidence of a pre-historic camp site at E.

xii. Carefully analyse, and record by field sketch and photograph, the view at F looking across Rhossili Bay to Rhossili Down. This involves descending to the bay to make a closer inspection of the staff section etc.

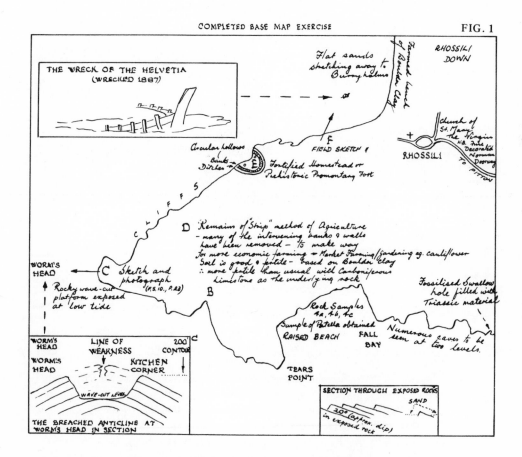

COMPLETED BASE MAP EXERCISE   FIG. 1

(f) *Sketch maps:* Older pupils should be encouraged to make notes in the field in the form of sketch maps. This provides excellent practice in landscape interpretation and

13

knowledge of map making. It also reduces cumbersome note taking. Hachures can be used for representing relief.

(g) *Making maps:* It is often necessary to make maps as a result of fieldwork. Distribution maps are quite commonly required, and these can be made by students at all levels from the primary school upwards by using the colour-headed map pins available from most stationers, or dots can be punched out of strips of coloured sticky paper and stuck on to the map. Fibreglass panels are very effective for making display maps, either drawn on to the panel, cut out of it, or by means of sticky paper stuck on to it. Completed fieldwork maps can also be drawn on to overhead projector transparency material, and these can be used either by the teacher or student to explain or demonstrate his work. A whole 'atlas' of fieldwork maps could thus be built up for projecting in this way.

2   Maps and Fieldwork in the Primary School

Maps are best introduced when the teacher is carrying out an environmental study of the school locality. Therefore the exercises suggested should be dealt with in conjunction with history, simple social studies and arithmetic, and when model making can go hand in hand with the whole project. If primary children are given the opportunity to work with maps they can quickly become quite skilled and avid map readers.

The maps which the children make themselves are most important in this learning process. The route from home to school can provide the starting-point. Home and school are drawn, copied from the blackboard, or duplicated copies distributed. The children fill in the details as they walk between these two points to make a one-dimensional straight line map as exemplified in Figure 2. Alternatively the two-dimensional route map can be used as in Figure 3.

FIG. 2

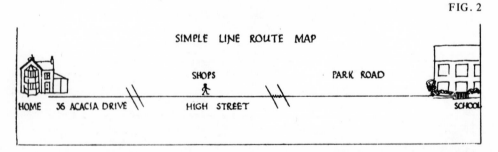

Another useful method is the map walk. In this the class is taken over a pre-planned circular walk. Many points of interest may be discussed en route but a minimum number, consistent with age and ability of the class, are pointed out for mapping later. After the walk, these selected features may be shown on a blackboard map and the children each make a free-hand map, adding their own extra details.

The question of symbols will arise at this point and this is often a suitable time to introduce some of the conventional O.S. map symbols. As an aid to learning, 'flash cards' of the commoner symbols can be prepared for use in the classroom.

14

FIG. 3

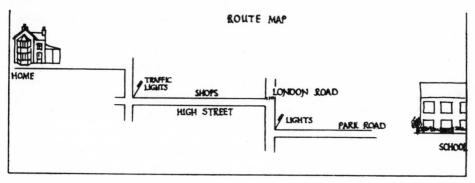

ROUTE MAP

3 Maps and Fieldwork in the Secondary School

The use of maps in the secondary school ranges from the first and second years when the children are developing a basic skill, to the more advanced use of maps as required for the fieldwork detailed in this book. In the latter stages, theory and practice can go more closely hand in hand — and there is still room for practical use of maps even at sixth form level.

Broadly speaking, in the first two years an awareness of scale, skill in using map references and the ability to orientate the map must be established. There is much overlapping here with work in the primary school and no hard and fast line can be drawn between the work in the upper primary school and the early years in the secondary school. Much depends on the ability of the children and the decisions made by the teacher.

(a) *Scale:* Children should have the opportunity to examine the 1″, 2½″, 6″ and 25″ to 1 mile O.S. map of their home locality. They soon appreciate the differences of detail shown on the various maps which vary according to their scale. Accurate practical knowledge of scale can be gained by making plans to scale of the classroom and then the school area. The distances of routes of imaginary journeys plotted on O.S. maps can also be computed in the classroom. Now is the time for attempting a sketch map to scale of the route from home to school. An appreciation of scale is further enhanced by the use of the viewfinder exercise. (See Chapter 4.)

(b) *Map references:* The practical approach to using map references can be made as follows:

i. A school with a large external area or field, or with a near-by park can try a 'map rally'. The first requirement is a base map of the area at a suitable scale and complete with a guide. Select 10 points for 'controls' represented by children standing at the appropriate place. Children are given the first map reference and proceed to the first control where they are given the next reference and so on. If the class is large, start the children from different controls so that each starting-point is set to make a 'circuit' in the correct order in a fixed lane. A further variation of this is to ask for certain selected fieldwork observations to be made at each control point.

ii. Where a large part of a morning can be set aside for this work and the school is in a country setting or in a school camp during the holidays, then 'map hunting' can provide a profitable and amusing exercise. This involves the children in learning how to orientate the map to landmarks in the landscape. It also involves the teacher in detailed preparation beforehand. The children work in pairs seeking hidden clues giving directions by map references. They have with them a tiny fragment of one of the O.S. maps which they are using (usually the route is made to cross over two or more sheets of the 2½" map). The hunt is up when they have arrived at the place shown in the small fragment of map issued them, and which they must identify by direct comparison of the map with the ground.

(c) *Orientation of the map with the ground:* This is best carried out on a short walk or traverse. While walking the map should be turned round so as to be able to read off the landmarks in relation to the position of the group — a simple skill to the initiated, but one requiring practice and understanding by the children if a map is to be used practically.

(d) *Using a compass to orientate the map:* The north point of the compass rose is made to coincide with the index-mark on the compass. The compass is then placed flat on the opened map in such a way that the marking notches on the lid (or the straight side of the compass) coincide with a True North and South line on the map. For this purpose the easting lines on the O.S. map can be taken as North/South lines.

The North and South of the compass dial now corresponds with the North/South line on the map. Rotate the map slowly with the compass upon it and watch the needle. When the dark end of the needle reaches the declination mark, indicating the direction of Magnetic North, stop turning the map which is now properly orientated with the ground. The landmarks can now be seen by the observer in the same relative position as on the map.

(e) *Past and present:* A useful elementary exercise is to provide pupils with large-scale maps of the school and its immediate surroundings as they were, say, thirty years ago. Duplicated copies can be issued to the children (the local library will probably have a suitable Ordnance Survey map as a source of information). The children are then sent out in pairs to record changes by drawing in freehand recent alterations to the map. Back in the classroom fair copies of the present-day map can be drawn and reasons for the changes suggested.

(f) *Morphological mapping:* The aim of morphological mapping is to add to the map by filling in significant details not shown on published maps.

Fieldworkers should remember that many contour lines on maps are really form lines plotted by interpolation. Details such as breaks of slope between surveyed contour lines or the position of the edge of a flood-plain are often highly significant, but they can only be discovered and plotted by the fieldworker.

Morphological mapping exercises should be confined to strictly limited areas. The detailed plotting of all the breaks of slope in one field may take a considerable time. As with all mapping one of the chief problems is to reduce data to manageable proportions. This type of mapping should always be carried out on a scale of 6" or 25" to the mile. Even on a 6″ map a line of width 1/50″ represents seventeen feet on the ground. Obviously it is not possible to map every small feature.

Much work on this type of mapping has been pioneered in the Geography Department at Sheffield University. Anyone interested in pursuing the study should read the articles published by the Geographical Association. One of these includes a very comprehensive key showing symbols which can be used for plotting many morphological features. It is advisable, with beginners, to choose a piece of open ground which contains varied topography or obvious breaks of slope. The area should be traversed along a series of straight lines in various directions. On each traverse all breaks of slope, angles of slope, direction of slope and distances between breaks of slope should be plotted (see Figure 4). They may then be recorded by drawing sketch profiles of each traverse, plotting symbols beneath to indicate the type of slope, etc. In this way pupils learn to devise and use simple symbols to plot morphological features.

IG. 4 TRANSECT ALONG BEARING 75 FROM THE N.W. CORNER OF NORLEY FIELD

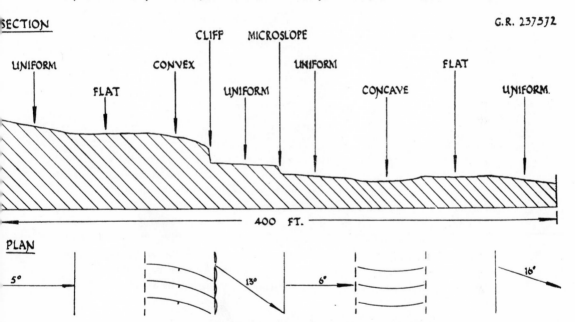

Solid lines show definite breaks of slope. Dotted lines show less significant breaks. Arrows show direction of slope. Curved lines show convex and concave slopes; an addition is made to the convex slope curve symbol to avoid confusion with the concave slope symbol.

4

The Visual Approach to Fieldwork

There are three basic methods of using a visual means for recording and evaluating fieldwork. These are: the viewfinder, the field sketch, and the photograph. In addition, there is a need to develop methods of evaluating the aesthetic quality of the observed landscape.

1   The Use of the Viewfinder

The purpose of using the viewfinder method is to give pupils practice in orientating maps and identifying landmarks, estimating distances, and measuring bearings.

To prepare a viewfinder the teacher needs to select a suitable viewpoint, locate it on the map, and then trace off lines radiating from it to obvious landmarks. True North should also be indicated. On a preliminary visit the card prepared in this way can be studied in relation to the actual view. It will be discovered that some landmarks cannot be used for this exercise as they may be obscured by trees and other obstructions. However, after having ascertained that the exercise can be carried out, a line is then drawn on the card to indicate the direction and distance to one obvious feature in the immediate foreground. This will enable pupils to orientate the viewfinder when it is in use. After the preliminary survey, the tested viewfinder can be duplicated for distribution to a class. Figure 5 illustrates a prepared viewfinder. When teaching beginners it is advisable to draw lines of a length proportional to the distances involved, but with more practised fieldworkers this will not be necessary and will be an additional test of their ability to estimate distances in the field.

On the site of the viewpoint the pupils set or orientate the viewfinder by the nearby object which has been marked in for this purpose. They can then check the alignments on the other landmarks shown. This is a useful preliminary exercise when showing pupils how to orientate maps. The distance can be estimated and pupils asked to suggest what scale can be used on the viewfinder. Protractors placed on the viewfinder may then be used to measure bearings: that is, the angle between True North and the object, measured in a clockwise direction. Results may be compared with compass readings, note being made of the magnetic variation. Eventually, those who have had some practice at this work may be encouraged to prepare their own viewfinders.

2   Introduction to Field Sketching

This is a complex technique which needs practising before a reasonable degree of competence can be achieved. It can be introduced to children by means of the outline field sketch, an example of which is given in Figure 6. This requires that the teacher prepares an outline field sketch of a view which is fairly familiar to the children. The drawing is then duplicated so that each child equipped with a copy of the outline field sketch can examine the view, identify the landmarks and complete the drawing for himself (See Figure 5).

FIG. 5 EXAMPLE OF A PRE-PREPARED FIELD SKETCH AS USED BY SECONDARY MODERN PUPILS

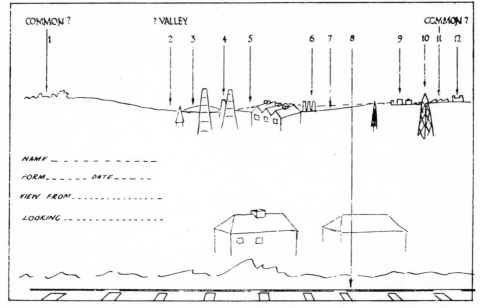

Classroom exercises can also be based on this work. Children can trace the main outline of landforms from selected photographs. They can, for example, look at a filmstrip photograph of a landscape view and practise sketching it. Alternatively the teacher can study a photograph with the class and build up a sketch on the blackboard. Thus by stages the pupil is introduced to the technique of sketching in the field.

3 Different Types of Field Sketch

(a) *The Detail Sketch:* it is often useful to make a sketch to record an observation that would otherwise take a long time to describe in words. For instance, a sketch of a geological section exposed in a quarry.

(b) *The Landscape Sketch:* this is the kind of field sketch generally attempted of a view overlooking an area displaying contrasts in topography, land use or settlement.

(c) *Skyscapes:* a view overlooking a town can be drawn with emphasis on representing the outline of the buildings, such as chimneys, church spires, point blocks, etc. as seen on the line of the horizon. This type of field sketch provides a relatively simple way of recording in visual terms the growth of a small town. Alternatively, skyscapes can be drawn looking across open spaces at part of a town or city's development. For instance, interesting skyscapes can be drawn looking across the River Thames to the buildings on the opposite bank of the river. If these are done at suitable intervals along

19

the river, and related to the map, they provide a visual record of the growth of London. A similar technique can be applied to the parts of an urban or industrial area opened up by a canal, such as the Manchester Ship Canal.

(d) *Comparative Field Sketches:* many old town maps show views or prospects over-looking the town at that time. It is highly instructive to make an attempt to compare such a prospect with a field sketch taken from as nearly a similar viewpoint as possible. Dr G. B. G. Bull in his article, 'Fieldwork in Towns — A Review of Techniques for Sixth Forms and Technical Colleges', *Geography*, Vol. XLIX, 1964, provides an interesting example of this kind of work attempted for a view overlooking Greenwich Palace and the Isle of Dogs, London. This article also illustrates the skyscape technique.

(e) *Panoramic Field Sketches:* reasonably competent pupils can make as a group project a panoramic field sketch. This is done by each person deciding on the section of the landscape view they are going to draw so that the three or more sections will, when drawn, connect up to make a wide view not obtainable by a single observer. These are then re-drawn, using a felt-tipped pen, on large sheets of cartridge paper and the whole drawing put together by pinning on to the classroom cork board. The panoramic effect obtained in this way provides an excellent visual aid to use in the classroom. In the process of its construction some accuracy might be lost, but this can be evaluated by active discussion centred round the drawing and related to the study of associated maps.

(f) *The Window Field Sketch:* this is for geography rooms having windows command-ing an interesting view over the school locality. By standing at arm's length from the window and using a felt-tipped pen, a fairly accurate representation of the view can be drawn on to the glass by tracing off the outline as seen through the windows. However, great care must be taken by the field sketcher in order to retain his original position during the process of drawing. If successfully completed this is a good method of bringing the view into the classroom. Added interest is gained by letting the outline drawn on the glass be projected on to a sheet of paper held before the window when the sun shines through it.

(g) *The Sketchmaster:* this is a development of the window field sketch, and is an aid towards better sketching in the fields. It is best used, however, for close and medium distance views. It is made as follows:
i. Hinge a piece of celluloid or perspex to the fieldwork board. This serves as a cover for maps etc. on which notes can be made, or sketches drawn, which bring out observed relationships to the topographical sheet below. It also provides protection from the weather.
ii. In order to make a field sketch the hardboard is rested on a firm surface, and the perspex raised at right angles to the board.
iii. Then view the subject to be drawn through the perspex, and without moving the head fix a pin into the nearside of the board. This will fix the sketch if it is lined up on a well defined object within the view to be drawn.
iv. Having lined-up the pin and object it is possible, as though looking through a window, to begin drawing on the perspex.

v. The main objective is to mark in the basic essentials of the view. So long as the head is kept fairly still it is quite possible to insert the major landscape elements without bothering about the technique of showing perspective, as this will come out correctly.

vi. Lastly, lower the perspex so that a piece of ordinary typing paper can be placed over it. Trace the outline on to the paper. This now forms the basis of a good sketch, and the remaining details can be completed by eye, and annotations made where necessary.

4   Making a Field Sketch

When making a field sketch the two main problems involved are the need to limit the extent of the picture, and to estimate proportions. Beginners can use a picture frame to help them limit the extent of the picture. Vertical proportions can be estimated by using a pencil held vertically at arm's length, the top of the pencil being sighted on the horizon and the thumb sighted on the selected base line. Horizontal proportions can be accurately determined by using a compass and taking bearings on important landmarks. It is wise to carry a supply of sharpened pencils, and the best for this purpose is medium soft. Sit down where possible to make the drawing and allow a fair amount of time for its completion if you are leading an excursion. Binoculars can be a useful aid to studying the view before drawing commences.

5   The Aims of Field Sketching

(a) The act of field sketching focusses the student's attention. He is made to look at the landscape, and in this way to improve his ability to observe in the field. A camera cannot do this for him.

(b) The field sketch is also a means of recording the analysis and interpretation of a landscape by the fieldworker. On the whole, artistic effects are not required, but honesty of vision is essential.

(c) The field sketch attempts to represent the basic structure (or 'skeleton') of the landscape, and is not meant to provide a naturalistic record of the view. Hence it is quite legitimate to use symbolic representations of landscape features, as shown in Figure 8.

6   The Method of Field Sketching

(a) Study the view using the Ordnance Survey map to identify landmarks and to make a first analysis of the landscape under observation.

(b) If possible, also compare the view with the geology map.

(c) Half-close the eyes and attempt to see the 'skeletal' structure of the landscape.

(d) As far as possible, decide upon a view that can be seen without turning the head.

(e) Then note down on the paper the map reference and description of the viewpoint, and the photograph number of the view (if taken at the same time). Also note the orentation of the view.

(f) Estimate the position of the horizon, relative to the centre of the page, and draw in as a faint line.

(g) Estimate the position of the foreground relative to the bottom of the page and draw in as a faint line. If there are a number of distinct areas within the view, delineate also with outlines (see Figure 6).

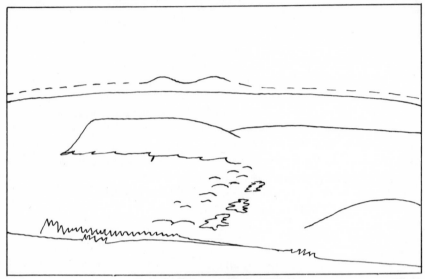

THE MAKING OF A FIELD SKETCH   FIG. 6

(h) Do not falsify distances, shapes, or positions of landscape features. If this occurs it is better to start again.

(i) Indistinct features can be shown as a broken line.

(j) On the whole, shading should be avoided.

(k) Slope can be indicated by means of perspective. Thus the tops of trees in the foreground Figure 7 get smaller as they descend the hill.

(l) Where telephone poles, houses, trees, etc. occur, these can be used for indicating scale as well as perspective in the drawing.

(m) Vegetation can be shown by means of symbols as shown in Figure 8. Other complicated features can also be shown in a similar symbol form. (See the sand dunes represented in Figure 7.)

(n) It is best to add settlement to the drawing after the basic structure of the landscape has been established.

(o) Make notes on the field sketch as you proceed. Not everything in the view is understood at the time of the drawing, and the notes should indicate observations that can be made to interpret the view, or that can be followed up later closer to the area in question.

22

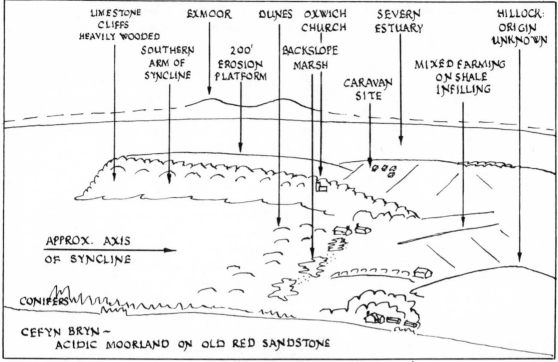

LIMESTONE CLIFFS HEAVILY WOODED

SOUTHERN ARM OF SYNCLINE

EXMOOR

200' EROSION PLATFORM

DUNES

BACKSLOPE MARSH

OXWICH CHURCH

CARAVAN SITE

SEVERN ESTUARY

MIXED FARMING ON SHALE INFILLING

HILLOCK: ORIGIN UNKNOWN

APPROX. AXIS OF SYNCLINE

CONIFERS

CEFYN BRYN ~ ACIDIC MOORLAND ON OLD RED SANDSTONE

FIELD SKETCH No. 16 FROM CEFYN BRYN (502895) (GOWER PENINSULA) LOOKING S.W. ACROSS OXWICH BAY SYNCLINE

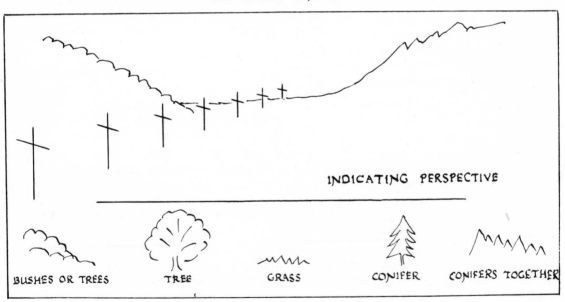

INDICATING PERSPECTIVE

BUSHES OR TREES   TREE   GRASS   CONIFER   CONIFERS TOGETHER

SYMBOLS FOR DRAWING LANDSCAPE FEATURES   FIG. 8

(p) The main method of annotating the field sketch in its final form is by means of carefully spaced vertical lines as shown in Figure 7. The annotations should not only indicate landmarks that have been identified, but in the case of advanced pupils should consist of interpretive comments too.

(q) Later, a finished quality can be given to the sketch by going over the pencilled lines with Indian ink and a mapping pen. But the accuracy and usefulness of the drawing is only achieved by careful observation in the field. The method outlined here is given as a guide for those who have little or no skill in drawing. Those who are able to progress further with the greater subtleties of field sketching should consult the late G. E. Hutchings' admirable book *Landscape Drawings* (Methuen, 1960).

7   The Use of Photographs

Photographs can play an important role in aiding the process of recording observations made in the field. Younger children will need guidance as to what photographs should be taken so as to derive the maximum value from the use of the camera. It is useful therefore to discuss with the children before setting off on the excursion the kind of photographs that would be worth taking. It is important that as the photographs are taken the fieldworker should record in his notebook the time, place and map reference. Once returned to the classroom, the successful photographs can be used to stimulate class discussion on the observations attempted. Transparencies are particularly useful for this. When photographs are finally mounted for display or put into the completed notebook they should be annotated in the same way as field sketches.

The colour transparency viewer used in the geography room is in many ways the equivalent to the microscope in the biology laboratory. Geography is taught as a practical subject; these 'geographical microscopes' need to be used more extensively than they possibly are at present. Analytical sketches similar to field sketches can be made of the view shown by the transparencies.

8   The Visual Evaluation of Landscape

In the various publications dealing with the methodology of geographical fieldwork there is little or no reference made to investigations which might lead the student to an aesthetic appreciation of the rural and urban landscape. It is true that this may be developed 'on the side', so to speak, whilst pursuing the more purely geographical types of enquiry. However, there is a danger that as geography teaching becomes increasingly influenced by the quantitative approach to the analysis of areal distributions now developing in the universities, the schools will not have time to develop an awareness of the quality of the landscape and environment. Professor Colin Buchanan once remarked, 'what we need is a cultural breakthrough in the importance of the physical environment'. Clearly, the schools should be at the spearhead of this cultural breakthrough, but the responsibility for this does not rest only on the shoulders of the geography teacher. For instance, the more creative attitude to art teaching which has entered schools since the war must give children a better chance to develop an aesthetic awareness, not only for paintings, but also for architecture.

24

That the young are not insensitive to the quality of their environment, once they are encouraged to observe, is shown by the recent activities of the Doncaster Junior Civic Trust when secondary pupils of all kinds proved equally capable at conducting street surveys. The purpose of their observations was to make suggestions to the local authority for improving the appearance of the town. This they did, and a number of alterations were made for the better.

There is certainly great scope for incorporating a study of the quality of the environment into a course on the humanities for the sixth form. This is because the evaluation of landscape reaches out into history, architecture, art and literature, as well as geography. The writings of Lewis Mumford, for instance, would be as cogent to this course as, say, D. H. Lawrence's descriptions of the Nottinghamshire landscape. The analysis of aerial photographs as well as fieldwork would also play a part. All that can be done here is to draw attention to a number of publications that are of value in working out the practical content of such a course.

(a) *The Changing Countryside Survey*. This is a scheme for involving schoolchildren in surveying the appearance of a selected area of the countryside at set intervals of time. Further information can be obtained from The Countryside Commission, 1, Cambridge Gate, Regent's Park, London N.W.1.

(b) *A Landscape in Distress*. This is the title of a book by Lionel Brett, published by the Architectural Press. It should be prescribed reading for all those interested in the quality of the landscape. It takes for its study an area of Oxfordshire and shows how modern suburbanisation and technology have eroded the beauty of this once rural landscape. The book is profusely illustrated with maps and photographs which themselves provide guidance for the way a similar investigation could be made by sixth form or college of education students.

(c) *Outrage*. This is by Ian Nairn and also published by the Architectural Press. Although an older book it provides many ideas for investigating the visual quality of the landscape by using map and photograph linked to the observations made on a traverse from Cumberland to Southampton. A similar but smaller traverse could be attempted by the older pupils.

(d) *British Townscapes*. A book by the geographer, Ewart Johns, and published by Edward Arnold. It seeks to explain the appearance of towns today by linking photographs, large scale maps and drawings to interpret the historical development of the townscape. In doing so it suggests many ideas for enquiry by older pupils wishing to investigate the quality of the urban environment.

(e) *Langstone Rock: an experiment in the art of landscape description*. This is the title of an article by Ewart Johns in *Geography* Vol. XLV, July 1960. In a very interesting preamble the author examines the concept of geography as art. This is followed by an attempt to portray in words the Langstone Rock, near Exeter. This is a kind of descriptive fieldwork exercise which can well be attempted by older pupils. Again there is an opportunity for comparison, and here Margaret S. Anderson's book, *Splendour of Earth* (George Philip, 1963) is a very useful anthology describing landscapes in many parts of the world.

5

Land Use Surveying

The aim of land use surveying is to classify and map the use of the land. Of all the techniques of fieldwork which exercise the pupil's map-reading ability none is more important than the detailed land use survey. The land use survey is also a valuable exercise in the recording of information in the field and it presents opportunities to practise plant identification. Besides having a social value and providing useful information, it serves to emphasise the importance of accessibility and demands a high degree of integrity from the surveyor, who must resist the temptation to guess at the use of a piece of land which is difficult to reach.

For school purposes the land can be classified in eleven major and two minor categories as shown in the chart.

	Land Classification Categories		
Land Use Group	*Colour*	*Derwent Crayon No.*	*Field Symbol*
Settlement	Grey	19–68	Pencil shading
Industry	Red	19–14	I
Transport	Orange	19–10	T
Public Open Spaces	Lime green	19–48	O S
Grass	Light green	19–46	G (GL if ley)
Arable	Light brown	19–61	A (additional letter to show crop)
Market Garden	Purple	19–23	M (additional letter to show type)
Orchards	Purple stripes	19–23	(tree symbols)
Woodlands	Dark green	19–45	(tree symbols)
Heath and Rough Land	Yellow	19–6	H
Water and Marsh	Light blue	19–32	(marsh symbol) for marsh / (symbol) for water
Derelict Land	Black dots		Pencil dots
Unvegetated	White		U

With only one deviation from the Old World Divisions of the World Land Use Survey – settlement – and with four additional categories – transport, open spaces, derelict land and unvegetated land – this scheme conforms with the New Land Use Survey. It does, however, avoid the fifty-two categories and additional data mapped on the New Survey, details of which are to be found in the Land Use Landbook. Though useful to advanced students, such a detailed survey is of little geographical value to most pupils as much time has to be spent in learning how and what to map. Once pupils have mastered the mapping of the thirteen categories suggested here, they may be encouraged to plot additional information to show the type of crop grown on arable land, etc.

In the field it is advisable to use a 6″ map. To avoid damage this may be covered with Kodatrace and recordings in the field may be made by writing the symbols on it using an HB pencil. If errors occur they can then be easily rectified, but care must be taken to see that the symbols are not smudged or unintentionally obliterated. When faced with a limited period of time the teacher could select a strip of land which crosses a variety of land use zones and is readily accessible from a road or footpath running like a spine down the centre of the strip. All the pupils have to do then is to plot the use of the land on each side of the road. If a class is divided into groups each responsible for its own strip, a considerable area can be covered and the data collected by the individual groups in a useful basis for comparative studies. Colouring-in of the maps should be done in the classroom.

When mapping has been completed, results should be carefully studied. In a country like Britain, where density of population is high, road improvements and housing development have encroached on agricultural land. The resulting conflicts between the various requirements of the community have to be resolved and pupils can find out which Authorities in fact control development in their area. This will help them to realise that in some areas political and social factors are more important than economics, e.g., land use may be controlled by legislation.

The most remunerative use of land often involves the investment of capital and this applies as much to agricultural land as it does in the town centre; the post-war farming revolution has shown that efficient application of capital can achieve spectacular increases in food production. Although much good farmland has been converted to other uses, farmers have got more out of the land already under cultivation and have reclaimed marginal land; e.g., heath, marsh and woodland. If pupils compare the results of their survey with the 1933 Land Use Map, they may be able to see where agricultural land has been taken for building schemes and where marginal land has been reclaimed for farming.

6

Examining the Soil

For the fieldworker the important consideration is that even on a local basis there will be great variety in soils in a very small area. Though local climatic differences are unlikely to be significant enough to affect soil, the varied combinations of relief, geology and vegetation will account for the surprising diversity of soils which await study in the local area.

1 Equipment

Some or all of the following equipment may be needed:
(a) spade and trowel
(b) soil auger. (This may be obtained from one of the suppliers of geological equipment, or may be made by a local blacksmith.)
(c) Soil testing equipment: i. 100 gms. barium sulphate (2s. 6d.); ii. 100 cc. B.D.H. universal indicator (3s. 3d.); iii. distilled water; iv. test tubes; v. B.D.H. colour card; vi. dilute hydrochloric acid; (All these are available on order from most local chemists).
(d) polythene bags or small tins for collecting samples.

2 Methods

Study maps of the locality and choose a number of sites where there are obvious environmental differences. Check that these are accessible and that property owners will not object to the digging of holes for the purpose of soil study. The steps to take when making a study of soil are set out below in the order in which they can be observed.

3 The Site

The points to be observed and recorded are: (a) pit, profile, or site number, (b) locality, (c) grid reference, (d) aspect, (e) slope, (f) drainage, (g) vegetation.

4 Colour

If the site is covered with vegetation a small patch should be cleared and a hole dug about one foot square so that students may look at the north face which will have the best light on it. In describing soil colour it is best to give a simple description (e.g., brown, grey, yellow) owing to the fact that light and moisture affect the colour seen. Colour is indicative of the mineral content: iron according to the degree of oxidation may give a yellow or red tinge, while calcium whitens a soil and humus darkens it. Waterlogged soil is often blue or grey with a mottled colour pattern. Colour will vary with depth, and samples of soil with obviously different colours may be collected in specimen tins for further study to discover any other differences.

5   Description

Classify the soil under the following headings:

(a) *Texture:* this is a physical property depending on grain size. It varies according to the proportions of sand, clay and humus which make up the soil. By handling, its classification can be determined, as shown in the table.

(i)	soil feels gritty	see *(ii)*
	soil not gritty	see *(iv)*
(ii)	soil form cohesive ball	LIGHT LOAM
	soil will not form ball	see *(iii)*
(iii)	soil stains fingers	HEAVY SAND
	soil will not stain	LIGHT SAND
(iv)	soil feels silky or sticky	see *(v)*
	soil not silky or sticky	LOAM
(v)	soil takes no polish	see *(vi)*
	soil takes a polish	see *(vii)*
(vi)	soil slightly silky	SILTY LOAM
	soil markedly silky	LIGHT SILT
(vii)	soil difficult to mould	CLAY
	soil not difficult to mould	see *(viii)*
(viii)	soil resistant but moulds	HEAVY LOAM
	soil only moulds with some difficulty	MEDIUM LOAM

Loam is the best medium for cultivating plants.

(b) *Constitution:* the compaction of the soil particles (interstitial spaces) may be expressed according to the way it handles and the way it appears.

i. Constituency (handling) express as:

loose: particles fall off auger or run through fingers.
compact: a good 'bite' with the auger, digs clean and well.
indurate: auger 'grinds' and spins, pick needed to dig.
friable: auger comes up loosely packed, digs nicely.
tenacious: auger 'sucks' and comes up full, spade clogs.
mellow: bores and digs well, ideal tilth, humus present.

ii. Porosity (visible) express as:

porpus: with many very small holes, 1–3 mm.
fissured: definite cracks, vertical.
spongy: rounded holes, various sizes 3 mm.
closed: no cracks or spaces apparent.

(c) *Structure:* the shape of the aggregate or nodules of soil may be expressed as:

crumb: roughly rounded (well-defined pore spaces in aggregate).
nutty: roughly rounded (auger — well-defined pore spaces in aggregate).
cloddy: conchoidal (ill-defined pore spaces in aggregate).
granular: solid rounded (ill-defined pore spaces in aggregate).
laminated: plate-like.

prismatic: may sometimes be jointed columns.
columnar: may sometimes be joined-up prisms.
pyramidal: peculiar structure of gleyed clays.
structureless: no aggregation of single particles.

A sketch can be drawn to show soil structure and the size of the aggregates should also be indicated.

(d) *Stones:* any coarse mineral skeleton material of dimensions greater than 2 mm. should be classified as stones and expressed in the following ways: adjectival description of chemical nature of stones – e.g., quartz pebbles whether of parent material or secondary origin.

quantity – express as: none, not frequent, many, few boulders, rock, rock dominant.

size – always given by name (e.g., coarse gravel) according to scale:

gravel	coarse gravel	very small stones
$\frac{1}{8}''-\frac{1}{4}''$	$\frac{1}{4}''-\frac{1}{2}''$	$\frac{1}{2}''-1''$

small stones	medium stones	large stones	boulders
$1''-2''$	$2''-4''$	$4''-8''$	$8''$

shape – expressed as: angular, sub-angular, rounded, shaly, tabular.

6   Moisture Content

Classify under the following headings:

(a) Organic matter: acting like a sponge, the organic matter on the surface and in the soil conserves moisture.
The degree of integration of organic matter with the soil can be indicated by reference to its nature, e.g., litter, raw humus, mild humus, intimate humus, mechanically incorporated organic matter.
(b) Roots: sketches can be drawn to show the root patterns.
The quantity, type and length of root will be closely related to the soil's moisture content. Roots should be considered under the following: quantity, size, health, age and shape or nature (woody, fibrous, etc.).
(c) Soil water: the degree of saturation may be observed by eye and by touch and expressed as: air dry, just moist, moist (colour changes on further wetting), wet (colour does not change on further wetting), waterlogged (water seeps and oozes).
Additional notes should indicate evidence of gleying (see profile diagrams, pages 33, 34.), height of water table and its fluctuations. On farm land record the method of drainage, e.g., mole or tile.

7   Fauna

Record evidence of the presence and activity of ants, worms, moles, rabbits, etc., particularly with reference to aeration and drainage of layers.

8   Soil Reaction

The fieldworker may test the acidity of the soil in three ways:

30

(a) Use of litmus. Moisten the litmus paper and smear with a soil sample. This is quick, simple but rather inaccurate way of determining whether soil is acid or alkaline.
(b) Use of hydrochloric acid. The application of a little dilute hydrochloric acid will indicate the amount of carbonates present in a given soil sample. The table opposite may be used to obtain a fairly reliable estimate.
The acidity of soil is measured by using a table of pH factors and hydrochloric acid can also be used as a rough guide to tell whether soil is acid or alkaline. A reaction will only occur if the pH factor is 7 or more, there will be no reaction if the pH is 7 or less. The pH value 7 is neutral, values above 7 are alkaline and those below are acid.
(c) Use of B.D.H. universal indicator. Distinguish between clayey, loamy and sandy soil, putting the appropriate amount of barium sulphate and then of soil in a clean dry test tube. The amounts for these soils are:

	Barium Sulphate	Soil
Clays	1½"	½"
Loams	1"	1"
Sands	½"	1½"

Percentage if Carbonates $(CaCo_3)$ present	Audible effects of reactions	Visible effects of reaction
0·1	None	None
0·5	Faintly audible, increasing to slightly	None
1·0	Faintly audible, increasing to moderate	Slight effervescence
2·0	Moderate to distinct; heard away from ear	Slightly more general effervescence
5·0	Easily audible	Moderate effervescence. Bubbles to 3 mm. easily visible
10·0	Easily audible	General strong effervescence. Bubbles to 7 mm. easily visible

Add distilled water. Shake well and allow to settle for a few minutes. Add reagent. Compare supernatant liquid with chart. The results are recorded as a pH reactor. The pH value 7 is neutral, values above 7 are alkaline and those below are acid.
There is an outer pH range beyond which some plants do not occur on a soil, and an inner range within which they flourish. A knowledge of the pH value enables the farmer to work out his lime requirements.

9 Soil Profile

This is the name given to a vertical section of soil. Each layer of the soil is known as a horizon and three main horizons are recognised — see diagrams which show the terms normally used. The profiles of soils are varied and the Figure 9a–d shows four typical profiles found in Britain.

Unless a natural exposure happens to be available, a soil pit has to be dug in order to study a profile. Exact instructions on the digging of soil pits are hard to give, but the following are points to bear in find:

(a) All horizons need to be exposed.
(b) The north wall should be as vertical as possible.
(c) The entrance should be stepped to provide ease of access and better light.
(e) There should be sufficient room for students to stand on the south side of the pit.
(f) A group of three or four can form a useful working party, taking it in turn to dig, record, use of equipment and carry out tests.
(g) A face two feet wide by two feet deep is more than adequate for soil study by school children.

10 Profile Characteristics

(a) These may be recorded on a scale drawing of the profile in a field notebook. Delimit the layers present from top downwards to parent material and for each record in the following order:

i. Depth in inches (indicate on scale profile).
ii. Clarity of definition of boundaries — express as:

sharply defined (i.e., change within 1″)   on scale profile
clearly defined (i.e., change within 2″)   on scale profile

iii. Lines on the scale profile may also describe the run of layers thus:

wavy ⌒⌒⌒, irregular ⋀⋁⋀⋁, smooth ―――― .

(b) Some people may prefer to use a standardised format for recording. One sheet has to be filled in for each pit, giving general information under the heading: Pit Number, Locality, Grid Reference, site aspect and slope, drainage, vegetation and tabulating the details under the headings: Horizons, Depth, Clarity of Definition, Colour, Organic Content, Roots, Texture, Size and number of stones — if any, water content and acidity as recorded by pH factor and Carbonate reaction.

(c) Another method of recording is by the construction of a soil monolith which is a scale model of the soil profile made from actual soil samples. The materials required are:

i. Soils samples should be collected in polythene bags from each of the horizons and carefully labelled.
ii. A strip of wood about three inches wide.
iii. Carpenter's glue and a brush.

32

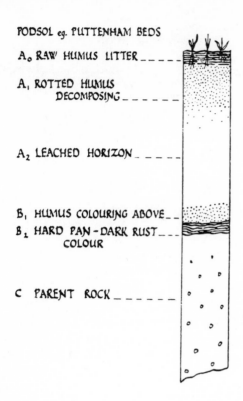

PODSOL e.g. PUTTENHAM BEDS

A_0 RAW HUMUS LITTER _ _ _ _ _

A_1 ROTTED HUMUS
        DECOMPOSING _ _ _ _ _ _

A_2 LEACHED HORIZON _ _ _ _ _

B_1 HUMUS COLOURING ABOVE _ _

B_2 HARD PAN – DARK RUST _ _ _
        COLOUR

C PARENT ROCK _ _ _ _ _ _ _

FIG. 9A

1. Profile sharply contrasted in horizons – top layer chiefly humus, leached layer ash grey, then humus darkening agove hard pan. Below the light-brown horizon grades into the parent rock.

2. pH value usually below 5·5; strongly acid.

3. Drainage excessive.

4. Occurs on sandstones in moderately heavy rainfall of temperate climates where precipitation exceeds evaporation.

5. The natural vegetation is heath.

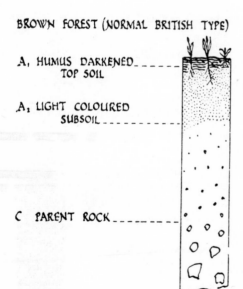

BROWN FOREST (NORMAL BRITISH TYPE)

A_1 HUMUS DARKENED _ _ _ _ _ _
        TOP SOIL

A_2 LIGHT COLOURED
        SUBSOIL _ _ _ _ _ _ _ _

C PARENT ROCK _ _ _ _ _ _ _ _

FIG. 9B

1. Profile more or less uniformly coloured, but with darkened humus, rich horizon on top.

2. pH value usually slightly acid and never base saturated.

3. Drainage varies from impeded to normal; only seldom is it free.

4. Natural (uncultivated) brown earths occur under deciduous woodlands in temperate climates. (N.B. Not a Podsol because drainage does not permit seepage of solutions to this extent.)

FIG. 9C

1. Single dark horizon is typical, but two are sometimes distinguishable (the top may grade into lower ones). Upper usually dark brown (but whitish on chalk). Grades into lighter colour below and fragments of parent rock are found here.

2. pH value usually 7 and above. Base saturated.

3. Drainage free.

4. Shallow soil, often on slope, with a characteristic vegetation of aromatic herbs. (N.B. Not a Podsol because chemical conditions counteract mineral loss through leaching.)

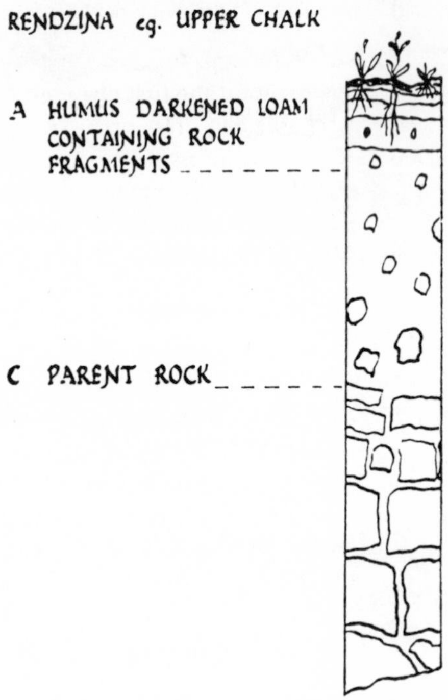

RENDZINA eg. UPPER CHALK

A HUMUS DARKENED LOAM
CONTAINING ROCK
FRAGMENTS _ _ _ _ _ _ _

C PARENT ROCK _ _ _ _ _

FIG. 9D

1. Waterlogged soil with anaerobic (airless) conditions preventing development of mature profile.

2. Glei horizon often over-lain by mottled horizon — sign of impeded drainage rather than actual waterlogging.

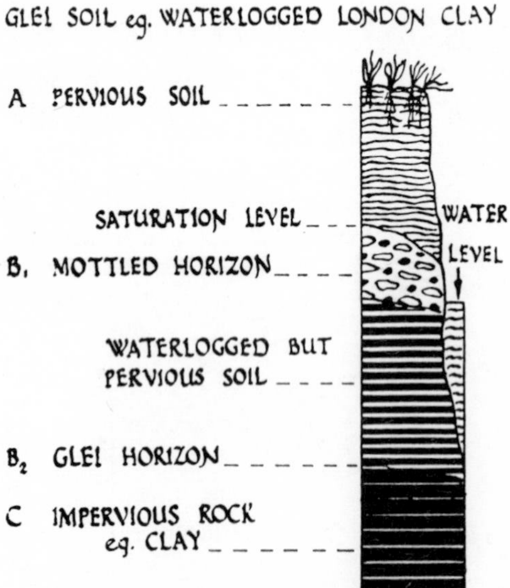

GLEI SOIL eg. WATERLOGGED LONDON CLAY

A PERVIOUS SOIL _ _ _ _ _ _ _

SATURATION LEVEL _ _ _   WATER
                            LEVEL
B_1 MOTTLED HORIZON _ _ _ _

WATERLOGGED BUT
PERVIOUS SOIL _ _ _ _ _

B_2 GLEI HORIZON _ _ _ _ _ _

C IMPERVIOUS ROCK
   eg. CLAY _ _ _ _ _ _

The monolith is constructed as follows:

i. Mark off horizons to a quarter scale on the strip.
ii. Prepare soil for the first divisions.
iii. Apply glue and press firmly on the soil. Turn over the board so loose soil falls off.
iv. Repeat the process for the other horizons, taking care not to leave any gaps. It is important to label the monolith, giving the following information: name and place of location, grid reference, brief description of exposure, height above sea-level, date, description of each horizon as seen in the field.

11   Use of the Soil Auger

It is inadvisable to use a soil auger until students have studied a soil pit. The auger should be inserted and twisted into the soil up to the shoulder of the bit — usually about six inches. It should then be withdrawn carefully and the core removed. Further samples may then be obtained from the same hole until bedrock is reached or the auger fails to penetrate any deeper. The chief value of the auger is that it can be used to obtain samples for testing soil texture and reaction. It is of particular value when quick tests are carried out during transects. It can also be used to find out how drainage and soil depth vary in different parts of a field.

12   Soil Fertility and Improvement

On agricultural land consider whether drainage is adequate, ascertain how frequently and how heavily fields are dressed with lime and find out the calcium oxide content of the lime. Other factors to consider are: depth of cultivation of the soil, frequency and timing of the application of organic manure, use of appropriate fertilisers, plant food deficiencies (nitrates, potash, magnesium and borun) and use of leguminous crops with their associated nitrifying bacteria. Kits, costing about one pound, can be obtained from reputable seed merchants to test soils for plant food deficiencies. Many of the more detailed tests referred to in this chapter can be carried out in the classroom using samples collected and carefully labelled in the field. Experiments can also be carried out in the classroom to show the percolation of water through soils of different kinds and the flocculating effect of lime. Reference may also be made to the hazards involved in the indiscriminate use of selective weed and pest-killing chemicals which may also kill birds and insects, the natural pest controllers and plant pollinators. A particularly useful kit for schoolchildren to handle is the Murphy Analoam Soil Tester, obtainable from the Murphy Chemical Company, St Albans, Herts.

13   Site Description

Field investigation of soil profiles should be accompanied by detailed descriptions of the site and profile. Field Study No. 20 in the Durham Field Studies Handbook for Teachers of Geography provides an example of two tables which can be used to summarise the main site and soil properties.

14   Soil Mapping

It is possible to proceed from this investigation to building up a rudimentary soil map.

In order to do this the investigation should take place where there are fairly contrasting soil types within a short distance. The sands, clays, and alluviums of the London Basin provide an example of such an area. A typical place within this area is Fairmile Common, near Esher, Surrey. Here there are several Tertiary sand and gravel areas near to the alluvium forming the floodplain of the River Mole. The method is for a group to work on the area covered by a 6" map. This is divided into equal sections among the group, and each section is then investigated for soil profile, site analysis, and dominant vegetation cover. The combined results of the group are then plotted on to a tracing paper overlay on the 6" map. It soon becomes apparent from such an exercise that soil profiles are not so uniformly developed as is represented in textbook diagrams. If care is taken in accurately recording the soil profile, and the area under investigation has been divided up into small enough sections, then it is possible to indicate changes in soil type. In the Fairmile area, for instance, podsols are found in the Bagshot Sand supporting a birch-pteridetum association, but the soils here are in various stages of development. There is a gradual change on to the Boyne Hill terrace gravels carrying an oak-heath association, and the soil profiles here approximate to a soil of the Brown Forest type.

15   Plant Geography

Once the student is reasonably skilful in soil analysis there is much useful fieldwork that can be done in plant geography. A traverse can be made across areas of contrasting soils, such as across a valley. At pre-determined intervals, soil samples can be obtained with a soil auger, and, also at pre-determined intervals, plants along the traverse line chosen can be identified. The result of this work is then correlated in the form of a transect diagram showing the botanical changes that occur, together with relief, geology, soil, drainage, and orientation of slope. Another example of this kind of work is the hedge analysis described on page 66 *et seq*.

7

A Farm Study

1   Aims

(a) To find out about farming. (b) To study the farmer's year — by visiting a farm in Autumn, Spring and Summer. (c) To try to understand the economy of a farm. (d) To use the study as a basis for comparisons with farming in other parts of the world.

2   Suitability

(a) Such a study is very suitable for an urban child. (b) This study is valuable to children of all ages, especially when related to work on the British Isles.

3   Equipment

(a) Topographical maps on 1″, 2½″, 6″ and 25″ scales. Geological maps on 1″ and 6″ scales. Soil-testing equipment. (b) Sheet to help in identification of crops and grasses.

4   Preparation by the Teacher

(a) Make contact with a farmer. (Apply to county branch of N.F.U. or better still, look for yourself). (b) Visit farmer and arrange visit, and whilst at the farm obtain background information — e.g., size of the farm, names of the fields, cropping plan for the year labour supply. (c) Duplicate base maps — 6″ to 1 mile is suitable as most farms at this scale will fit an exercise book. (d) Prepare questionnaires.

5   Preparation by the Children

(a) Study O.S. map to locate farm. (b) Colour base maps according to cropping plan. (c) Draw diagrams to illustrate crop rotation (using same colour key as in (b)).

Sample Study

Felton's Farm (about 2 miles S.E. of Dorking), Mr L. Trick.

1 *At the Farm*
(a) Locate farmer and make introductions. (b) Compare base map with landscape. (c) Note farm buildings, their layout and uses. (d) Note age of buildings, kind of building material. (e) Note sources of power, water supply and system of drainage.

2 *Potatoes*
What to note:

(a) Autumn
i. Progress of potato harvest, ii. how a clamp is made, iii. estimate of crop (if a clamp

is 2 yards wide, every 2 yards of length = 1 ton), iv. size of riddle for the year decreed by Potato Marketing Board, v. dyeing of little potatoes purple (not to be sold in shops) – used as pig food.

(b) Spring
vi. Number of acres sown, vii. amount of government levy per 'quota acre', viii. amount of government levy per acre over 'quota acres'.

(c) Summer
ix. Ridging of crop, x. trial patches of fertiliser (Fisons).

(d) General xi. That this is a cash crop, xii. kind of potatoes grown, xiii. how often fresh seed potatoes are purchased from Scotland.

3 *Other Root Crops* (swedes, mangolds, marrow-stemmed kale)

(a) Autumn
i. Size of each crop, ii. how and where stored, iii. any vulnerability to frost, iv. deterioration with age.

(b) Spring
v. Size of seed, vi. whether plants are showing yet.

(c) Summer
vii. Appearance of crop.

(d) General
viii. Use of crop (cash, stall feeding), ix. marrow-stemmed kale is not really a root but is grown in the field devoted for the year to roots. It can be used for strip grazing controlled by an electrified fence.

4 *Grass*

What to note:

(a) Autumn
i. State of pastures, ii. new pastures – undersown?

(b) Spring
iii. Fields reserved for hay.

(c) Summer
iv. Progress of haymaking, v. storage of hay.

(d) General
vi. Use of hay – stall feeding.

5 *Water Supply*

What to note:

(a) Autumn
i. Type of 'cup' used for stall feeding, ii. how Tanner's Brook is dammed to give a head of water to work Blake's Hydram, iii. how Blake's Hydram works, iv. how water is available from troughs in each field and filled by hydram.

6 *Sheep*

What to note:

(a) Autumn

38

i. Size of flock and breed, ii. how recently purchased, iii. that they will stay in the open throughout winter.

(b) Spring
iv. How many fattened sheep are being sold each week.

(c) Summer
v. No sheep left on farm.

(d) General
vi. That sheep are 'scavengers' — eat closer into hedges, nibble grass shorter, tread new pastures firmer without damaging them.

7 *Calves and Steers*

What to note:

(a) Autumn
i. Calves are shut up and stall fed for first six months oof their lives, ii. whether steers are shut up yet.

(b) Spring
iii. That calves are blinded by light when let out and bump into obstacles.

(c) Summer
iv. Varying ages of different groups of steers in different fields.

(d) General
v. That by staggered buying the age of the groups will vary and cash from sales be less spasmodic, iv. the breed (Devon × Hereford), vii. subsidy paid by government on beef cattle but paid once only (hole punched in ear), viii. dehorning — electric or caustic soda, ix. markets used, x. 'diet' for stall feeding, xi. accumulation of muck.

8 *Degree of Mechanisation*

What to note:

(a) Autumn
i. How a potato lifter works, ii. use of tractor and plough.

(b) Summer
iii. Use of haymaker, iv. use of elevator, v. how a binder works, vi. how a combine harvester works, vii. how a baler works.

(c) General
viii. Whether machinery is owned or hired.

9 *Wheat, Oats and Barley*

What to note:

(a) Autumn
i. Whether any of above have been sown yet.

(b) Spring
ii. Resemblance to grass.

(c) Summer
iii. Shape of ripening ears of each crop, iv. that breeds of each of these crops have names and more recently numbers, v. uses of each crop, vi. that a good field with few rogue ears earns a certificate and commands a higher price for seed than for milling, vii. uses of straw — litter, some can be eaten by stock and for thatching, viii. that

39

FIG.10

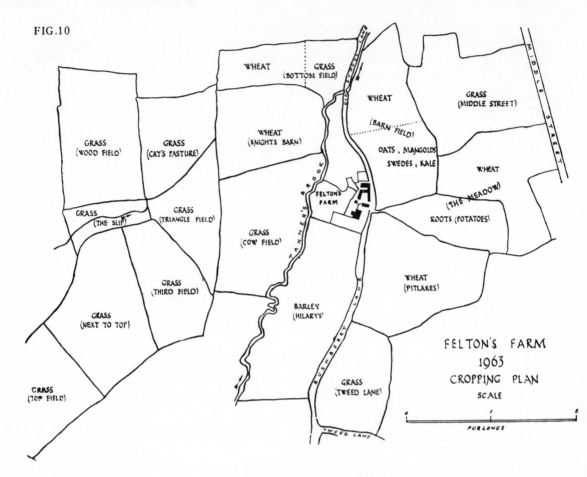

WHEAT   GRASS
(BOTTOM FIELD)

WHEAT

(BARN FIELD)

GRASS
(MIDDLE STREET)

WHEAT
(KNIGHT'S BARN)

OATS, MANGOLDS
SWEDES, KALE

WHEAT
(THE MEADOW)

GRASS
(WOOD FIELD)

GRASS
(CAT'S PASTURE)

FELTON'S
FARM

ROOTS (POTATOES)

GRASS
(THE SLIP)

GRASS
(TRIANGLE FIELD)

GRASS
(COW FIELD)

GRASS
(THIRD FIELD)

WHEAT
(PITLAKES)

GRASS
(NEXT TO TOP)

BARLEY
(HILARY'S)

FELTON'S FARM
1963
CROPPING PLAN
SCALE

GRASS
(TOP FIELD)

GRASS
(TWEED LANE)

0   1   2

FURLONGS

TWEED LANE

FIG. 11

ROTATION OF CROPS

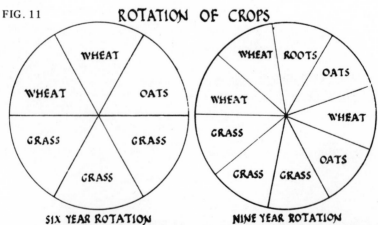

WHEAT

WHEAT   OATS

GRASS   GRASS

GRASS

SIX YEAR ROTATION

WHEAT ROOTS

OATS

WHEAT

WHEAT

GRASS

OATS

GRASS GRASS

NINE YEAR ROTATION

Note.–Topics are dealt with in the order they are likely to be encountered during the 'Farmers' Year', i.e. activities seen during Autumn onwards.

these are largely cash crops, ix. that long straw from binder is fancied by horse breeders and is in short supply (combine harvesters spoil straw as far as horses are concerned), x. that straw for litter, with stall feeding, affords manure for spring muck spreading, xi. how cattle rearing and corn growing 'dovetail' together.

10 *Markets*

What to note:

(a) Names of nearest markets – Dorking, Guildford, Horsham. (b) Wheat, oats and barley are not taken to market but sold by sample. (c) Sheep grading at markets – sold in groups to butchers. (d) Steers sold individually at markets to butchers. (e) Other things sold at market.

11 *General Notes*

(a) This farm is visited three times a year. (b) Although most people tend to regard the farmer's year as starting again after the corn harvest, some crops are still being harvested, e.g., roots, in autumn. (c) According to the capabilities of the children taking part, work can be undertaken on soil texture and pH values, etc. (d) On the summer visit, the first call is at Dorking Market (Mondays) where the two auctioneers who alternate (Mr Crow and Mr Pringle) welcome the children and explain the animated scene.

6 Further kinds of Investigation

Comparative Studies
Divide a class of older pupils into groups, so that each group makes a study of a farm within a certain area.

Land Use Maps: the land use map of the farm compiled during fieldwork can be compared to the land use of the same area represented on the 1930's land use map. In some cases, crop returns for much earlier years are also available in the County Records Office, and interesting comparisons can be made between these and the agricultural statistics available from the Ministry of Agriculture, London Road, Guildford, Surrey.

The Farming Year: in the form of a divided circle, compile a diagram to show the activities on the farm during the different seasons in the year.

Site	Altitude, aspect, slopes.
Access	Cart track, lane, classified road, railway siding.
Fields	Sizes, access fields, material of boundaries.
Soils	Depth, drainage, erosion, fertilisers, pH factor, liming, reclamation and improvement.
Grazing	Kinds of grasses and grass mixtures, rough grazing, grazing off the farm, areas never ploughed, areas reverting to natural vegetation.
Crops	Sources of seeds, 'look' of seeds, rate of sowing, seed dressings, yields, double cropping, catch crops, silage.
Stock	Beef *v.* milk, milking times, milk yields, town

	milk supply, dried and processed milk products, butter and cream, calves per cow, frequency, imported foodstuffs, cattle licks, breeding policy, A.I., dehorning, milking parlours, loafing yards. Transhumance of sheep.
	Size of litters (pigs), frequency, breed, baconers or pork.
	Hens, open range, deep litter, battery.
Subsidies	Levies and policies of the Marketing Boards.
Tenure	Tenant, owner.
Labour	Regular, casual, students and pupils, their training and accommodation.
Mechanisation	Modern implements, cost, independence from weather conditions thereby.
Water supply	Buildings — piped, well, stream, 'cold bank' for dairy. Fields — dewponds, ponds, piped, irrigation.
Electricity	Generated — wind, water, oil engine. Mains.
Buildings	Old and new, subsidised or not, extent and character, function, adaptation with changes of method or policy.
Pest Control, Diseases	Rats and other vermin, foot-and-mouth disease, ravaging by dogs, diseases completely eradicated, sheep dips, the newer 'nervous' diseases in cattle. Crushing of sucking-pigs.
Veterinary Services	How easily obtainable (nationalisation?).
Marketing	Local and distant. Open market v. direct sales.
Repairs	Welding and spare parts for machinery.
Shooting rights	Owned or let, 'bag'.
Insurance	Against weather, fire, etc.
Experimental work	Co-operation with local Farm Institute, Ministry of Agriculture, University Department.
Modern trends	Specialisation, vertical integration, horizontal integration.
Prospects	European Common Market.

8

The Traverse

The traverse is a basic technique of fieldwork. It provides a simple orderly framework on which to 'hang' the observations made during the excursion.

1   The Traverse with Transect

The method consists of taking as straight a course as possible across the terrain to be studied, and recording selected data seen on either side of the line thus followed, e.g., the Effingham to Wotton traverse. This technique is best applied in country or town where the route chosen traverses belts or zones of considerable geographical contrast. It is wise before setting out to decide on the kind of observations that are to be made. To attempt to record everything could result in confusion.

The pupil should have, and understand, a base map of the geological formations over which the traverse passes. This can be simplified for the younger or less able child provided the location of the traverse is suitably chosen. The more mature student should prepare the base map for himself.

Each child must have a transect diagram. This is prepared by making first a cross section from the relevant O.S. map. A profile of the rock strata outcropping along the line of the traverse is drawn in the correct position and scale. Above the transect diagram boxed spaces can be left for recording selected observations. The transect diagram used for the Wimbledon Common excursion is an example (Figure 12).

In the field, with the aid of the base map and transect diagram, the pupils can relate changes in land use, natural vegetation, soil, drainage and settlement to the changes in rock formation over which the traverse passes.

The observations recorded in the box spaces can be made in longhand or by means of a series of symbols agreed to before setting out.

Returning to school, the full synthesis is made by the children constructing an enlarged and accurate transect diagram suitable in size for classroom display. The pupils contribute their own observations made in the field, placing them under the different categories. This allows for a more comprehensive record being made, and gives point to follow-up discussion, as not all the children will necessarily have a complete record in their field notes. This method is most useful when the transect is being completed for the first time. If the completed diagram is pinned on to the wallboard then fossil, plant and soil specimens, etc. collected en route can be displayed below and alongside the appropriate point where they were discovered during the making of the transect.

FIG. 12

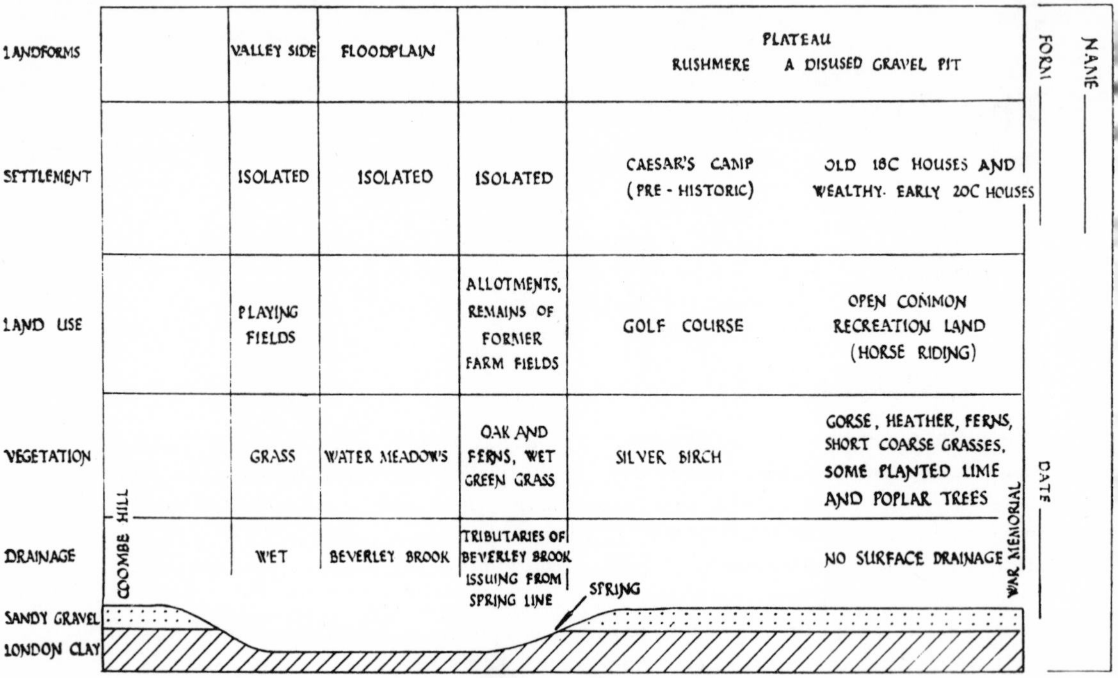

LANDFORMS		VALLEY SIDE	FLOODPLAIN		PLATEAU — RUSHMERE A DISUSED GRAVEL PIT		FORM _____ NAME _____
SETTLEMENT		ISOLATED	ISOLATED	ISOLATED	CAESAR'S CAMP (PRE-HISTORIC)	OLD 18C HOUSES AND WEALTHY EARLY 20C HOUSES	
LAND USE		PLAYING FIELDS		ALLOTMENTS, REMAINS OF FORMER FARM FIELDS	GOLF COURSE	OPEN COMMON RECREATION LAND (HORSE RIDING)	
VEGETATION	COOMBE HILL	GRASS	WATER MEADOWS	OAK AND FERNS, WET GREEN GRASS	SILVER BIRCH	GORSE, HEATHER, FERNS, SHORT COARSE GRASSES, SOME PLANTED LIME AND POPLAR TREES	DATE _____
DRAINAGE		WET	BEVERLEY BROOK	TRIBUTARIES OF BEVERLEY BROOK ISSUING FROM SPRING LINE SPRING		NO SURFACE DRAINAGE	WAR MEMORIAL
SANDY GRAVEL							
LONDON CLAY							

COMPLETED TRANSECT DIAGRAM OF THE WIMBLEDON COMMON TRAVERSE (AS COMPLETED BY A 2ND YEAR B STREAM MODERN PUPIL)

2   The Simple Traverse

This is a simple method of recording observations made along a route chosen for its geographical interest. A study of Figure 13 shows how a base sheet should be provided for each student who then proceeds along a route indicated to him on the relevant O.S. map. As is shown in Figure 13 the record is made to right and left of the route including only as much as can be seen from the road. This method lends itself to pupils of all ages. It does not have to follow a straight traverse as in the traverse with transect described above. The simplicity of this method should not be allowed to obscure the fact that this is one of the most useful ways of conducting school fieldwork.

The observations made on the traverse can then be transposed on to a base map drawn from the 6″ or 2½″ O.S. map of the route taken. Again this can be enlarged sufficiently to make it of a size suitable for classroom display and assembly in groups. This also allows for field sketches, photographs and specimens to be arranged round the completed base map.

44

FIG. 13

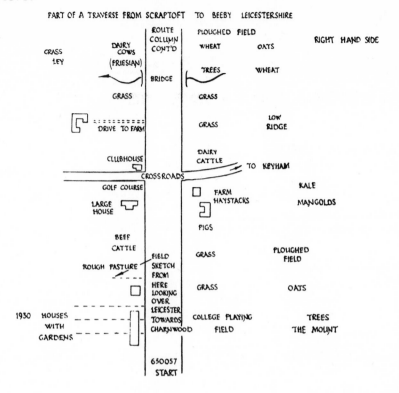

PART OF A TRAVERSE FROM SCRAPTOFT TO BEEBY LEICESTERSHIRE

45

9

A Traverse with Transect across an Escarpment

Effingham to Wotton

1   Aims

(a) To demonstrate the land utilisation and vegetation change with the changes in the underlying rock.
(b) To demonstrate the physical geography of a chalk escarpment.

2   Suitability

This is perhaps the simplest possible traverse across the North Downs. Wisely presented, it is suitable for all ages.

3   Time

4 hours, including lunch break.

4   Distance

4 miles.

5   Equipment

1″ O.S. map sheet 170. 2½″ O.S. map sheets TQ 14 and 15. Geological Survey Map, Reigate Sheet. Geological hammer. Soil auger. Clinometer. (The last three only if available.)

6   Preparation

(a) *Information*

A preparatory knowledge of the London Basin and the Weald.

(b) *Activity*

Study the route on the O.S. maps; prepare a contour map extract; study geological map, transect diagrams and model of the area to be traversed.

7   Route

Turn right outside Effingham Junction Station, walk along road crossing Effingham

Common, in times past an intractable area given over to pannage and hence commonland.

What to note: i. the crest of the North Downs to the south, ii. Characteristic flat aspect of the common area. iii. Large puddles after rain typical of claylands; also small streams draining away surface water, e.g., 110550. The surface is cracked and parched in the heat of summer. iv. This is at the northern boundary of Effingham Parish.

Activity: i. compare geological base map with the ground. ii. Record land use and vegetation – mainly oak and mixed farming on clay soil. (i. and ii. should be done throughout the excursion.) iii. Collect a ball of clay in the hand, roll between fingers, notice typical clayey texture. iv. Photograph the featureless aspect of Effingham Common. (To record views of geographical interest, photographs should be taken throughout the excursion.)

8   Route

Continue along road to Effingham.

What to note: exposure of the Taele gravel in road cutting at 125546. The flint stones are large and irregular – not water worn – and are embedded in a red sandy matrix. This is sludge residue from the North Downs deposited by the process of solifluction. During the Ice Age this area was in the Periglacial or Tundra zone.

Activity: i. collect and label specimen of flint gravel. ii. Photograph exposure.

9   Route

Pass through Effingham Village. This could be made the subject of a detailed study. The suffix -ingham indicates it was a Saxon settlement originally. It is a spring-line village.

What to note: i. small fields of irregular shape near to the village, possibly the product of early enclosure. ii. Advanced students should look for evidence of the narrow outcrop of Reading and Thanet sands which occurs here, between the traffic lights and the golf course. iii. Use of flint as a building material. iv. The milk marketing centre in the village, built since the last war and collecting from the dairy farms on the local claylands. v. New estate of 'commuter' housing.

10   Route

Continue through village, then take the first left-hand fork in the road. This begins ascent of the dip-slope.

What to note: i. the chalky soil. ii. The marked change in vegetation and land use. Beech and yew now very prominent, the large fields are either under the plough for cereals or ley grasses. These very extensive fields are the product of post-1914 enclosure. iii. The way the road runs on an interfluve between dry valleys.

Activity: i. identify crops growing hereabouts. ii. Estimate the gradient of the dip-slope. iii. At about 120526 look northwards over London Basin. Set maps. Identify major landmarks such as St. George's Hill; Weybridge; Chobham Heights; Esher Common; and larger buildings of London, e.g., Senate House. Use 'viewfinder' method for recording this observation.

11 Route

Continuing along road at 122578 pass down into Polesden Lacey Valley. This runs in the strike direction of the chalk escarpment.

What to note: i. the dip-slope shows a general absence of surface water. ii. The presence of dry valleys.

Activity: this is a good opportunity for studying the cross-section of a valley. By simple survey methods construct an accurate cross-section diagram of the valley.

12 Route

Continue to 119501.

What to note: i. change of soil here because chalk is masked by clay with flints. ii. The fact that dry valleys here do not notch the crest of the escarpment.

13 Route

Turn left at the road junction, walk eastwards along road. Opposite first house on the left, climb the gate and take the right hand of the two bridle paths leading through the woodland to White Down. This is at the southern boundary of Effingham Parish which is thus a typical 'strip parish'.

What to note: i. beech trees with undercrop of poorly developed oaks growing on clay with flints. ii. Pickett's Hole (121494). Looks like a chalk pit or bomb crater; more possibly a solution hollow in the chalk. iii. White Down (122494). The scarp face of the chalk escarpment is revealed here. Study the surf and chalk-loving flora, the beeches and yews typical of downland vegetation. iv. The characteristic buttress formation of scarp face. v. The opening in the trees giving a fine view over the Vale of Holmesdale and the northern edge of the Weald. The railway line at the scarp foot runs on the Upper Greensand bench.
WHITE DOWN. (This is a good resting-place for picnic lunch.)

Activity: i. set maps. Identify distant and near landmarks – e.g., Dorking, Box Hill, The Nower, Leith Hill, Vale of Holmesdale, Deer Leap Wood, etc. Record by means of 'viewfinder' method. ii. Draw field sketch (or photograph looking east towards

48

Dorking), to show scarp face and Vale of Holmesdale. iii. Estimate gradient of scarp face.

14 Route

Follow footpath down scarp face.

What to note: small chalk pit (halfway down) dug into Middle Chalk.

Activity: Collect ammonites, pectines, inoceramus fossils and hand specimens of Middle Chalk. Label these.

15 Route

Follow footpath over Upper Greensand bench, crossing railway bridge.

What to note: i. downwash from chalk overlapping on to cultivated fields situated on fertile Upper Greensand. ii. Rapid change of vegetation from scarp face to Upper Greensand bench. iii. Narrowness of Gault Clay outcrop. Advanced students would try to locate this using a soil auger.

16 Route

Continue along footpath through oak woodland.
What to note: i. abrupt change from the Upper Greensand and Gault Clay belt to the Lower Greensand, marked by mixed woodland (mostly oak,) with rich undergrowth of bracken growing on Lower Greensand outcrops. ii. Sand pit in Folkestone sand at 127485. This shows current bedding and banded layers of carstone.

Activity: i. examine, sketch and photograph the current bedding. ii. Collect and label specimens of carstone.

17 Route

Continue along footpath to Dorking/Guildford road. Where footpath joins road are bus stops for buses to Guildford or Dorking.

10

Fieldwork from the Air

The number of schools chartering aircraft for a short educational flight has increased enormously over the past few years. Such excursions, if part of a carefully thought-out scheme of lessons, provide a most worthwhile stimulus to learning. This is a situation, par excellence, where children can obtain a direct and vivid experience of the world beyond the classroom. Certainly as they fly over the countryside and the town, the map is brought to life and a new perspective is gained of the environment in which they live.

The major difficulty in organising such a venture is caused by the uncertainty of predicting favourable weather on the day of the flight, though given reasonable conditions all should go well. Of equal, if not greater importance, is the careful consideration of the methodology of this novel kind of fieldwork.

1 Pre-flight Preparations

Meticulous preparation and organisation is essential. The following points should be taken into consideration:

(a) *Finance*: It is not possible to state actual costs as these will vary from the time of writing. But obviously this is an expensive form of travel and the teacher would do well to organise the collection of the required money over a period of time before the date of the flight. It is also important at this stage for the teacher to investigate the question of insurance cover with the Local Education Authority and the air charter firm concerned.

(b) *Consultation with the air charter firm:* It is invaluable to enlist the co-operation of the charter firm. The teacher must be able to evaluate from the advice he obtains what he can get from a flight of a particular duration. Half an hour might be considered the optimum period for a study flight. If there is any choice possible, then an aircraft like the Dove is preferable, say, to a Dakota because the former has the best windows for viewing. Some schools have even used helicopters. Other points to consider are:

i. The speed at which the aircraft will travel. Upon this will depend the distance that can be covered in a limited time.

ii. The height the aircraft will fly. Fifteen hundred feet is the regulation lower limit.

iii. Ensure that the pilot is willing to make turns over particular areas chosen by the teacher. Usually this is possible. In the case of the Dakota, each turn takes two minutes flying time approximately, and about four minutes for a figure-of-eight turn. If turns are required then these must be allowed for when estimating the distance to be covered in the time available.

iv. How to indicate the route to the pilot. Pencilled directions on a transparent overlay covering the area on a relevant 1″ Ordnance Survey map is useful for this.

vi. The use of cameras and portable tape recorders. Usually permission can be obtained for the use of these. Although the background noise of the aircraft can mar the quality of the recordings, it is useful (and sometimes entertaining) to record immediate impressions for playback in the follow-up work to the flight. But it is more important that quite ordinary cameras can obtain useful black and white photographs of the view from the aeroplane, provided the weather conditions are reasonable. In the case of colour transparencies, an ultra-violet filter should be fitted to the camera, and a meter reading taken from grass or buildings just before take-off according to whether the flight is over rural or urban landscape. The shutter speed should be taken down to 1/100th second, or better still, to 1/250th second. The best pictures are obtained with the sun over the left or right shoulder, but allowances have to be made for changing light conditions. In any case, one must be prepared for a certain amount of wastage of the film, as the irregular movements of the aircraft, and possible dirt on the windows will also affect results.

vii. The seating plan of the aircraft. This is useful to know so that children can be allocated their seats beforehand. This will also facilitate an orderly entry into the aircraft when the children may be excited, and somewhat apprehensive, if they have not flown before. In the Dakota, the five pairs of seats at the rear give a view clear of the wings.

viii. Visiting the pilot's cabin during the flight. If permission is obtained for this another interest is added to the flight.

ix. Alternative dates for the flight. These should be arranged with the firm in the event of cancellation due to bad weather.

2   Determining the Flight Route

This will be consistent with the aim of the preparatory work completed in the classroom. It is important to take into consideration that the children being unused to flying will find themselves disorientated by the novelty of the occasion. They have to adjust to seeing the landscape passing below, from an unfamiliar viewpoint, and restricted by the narrowness of vision from the aircraft's windows. It is useful, therefore, if the route can follow the line of a river, or cross over contrasting areas and landmarks which can be readily identified.

3   Suggestions for Pre-flight Preparations with the Pupils

As with all fieldwork, the air excursion should be closely integrated with the scheme of work being followed by the pupils. The preparation will depend on the attainment and age of the children concerned, as well as that aspect of geography which they have studied prior to the flight. Here are some suggestions for preparation which could be undertaken with a non-specialist class:

Work out with the class the possible route the aircraft might take; develop case studies based on mapwork of the contrasting areas which will be seen from the air. The pupils can also make sketch-map studies of these areas with particular reference to land uses, communications, land forms, and settlement. It is particularly valuable for the children to visit part of the area over which the aircraft travels; field study

FIG. 14

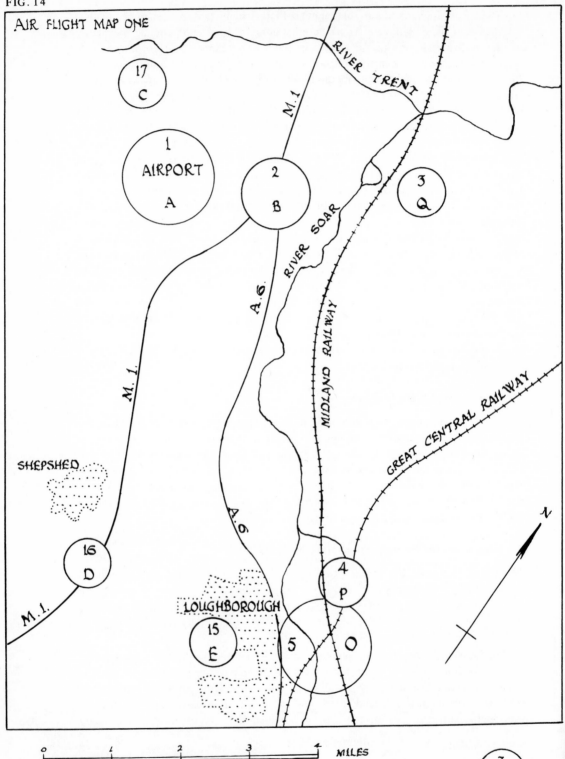

AIR FLIGHT MAP ONE

RIVER TRENT

M.1

17
C

1
AIRPORT
A

2
B

3
Q

RIVER SOAR

A.6

MIDLAND RAILWAY

M.1

GREAT CENTRAL RAILWAY

SHEPSHED

16
D

A.6

M.1

4
P

LOUGHBOROUGH

15
E

5

O

N

0 1 2 3 4 MILES

URBAN AREAS

NOTED FEATURE

3
Q

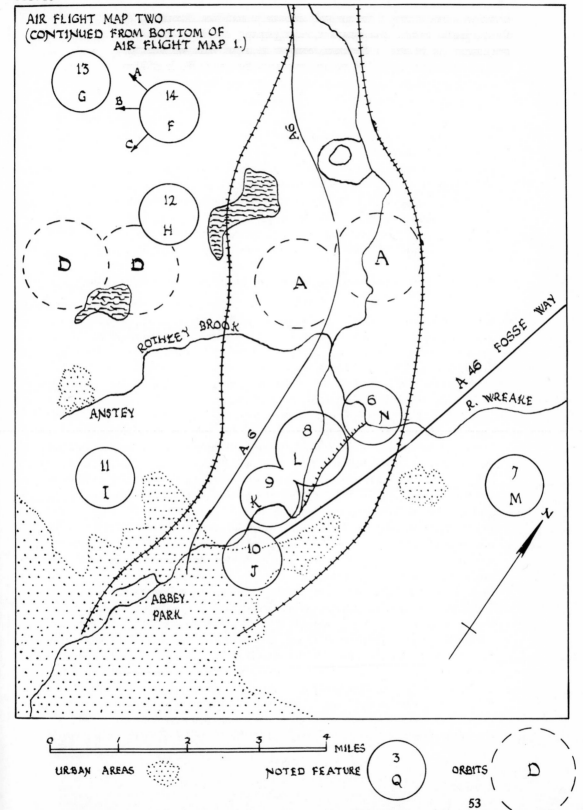

FIG. 15

AIR FLIGHT MAP TWO
(CONTINUED FROM BOTTOM OF
AIR FLIGHT MAP 1.)

13 G

14 F

A

B

C

12 H

D

D

A

A

A6

ROTHLEY BROOK

ANSTEY

6 N

8

A6

L

11 I

9 K

A 46 FOSSE WAY

R. WREAKE

7 M

10 J

ABBEY PARK

0 1 2 3 4 MILES

URBAN AREAS

NOTED FEATURE

3 Q

ORBITS

D

53

excursions to one or more of the case study areas can be useful here. Alternatively, if it is not possible to make these excursions, then a 'geography from the coach window' questionnaire can be used on the coach journey to the airport which will direct the children's observation to part of the area over which they are to fly. In addition classroom projects could be followed prior to the flight, and these could allow for team teaching, or a combined study approach. For instance:

(a) The development of air travel. This could include the history of flight, the scientific theory of flight, the geography of air transport, a visit to an aircraft factory, and a study of a local airport.

(b) The world from the air. This would entail the use of air photographs illustrating not only the local area, but also contrasting environments elsewhere. C. F. Casella and Co. Ltd. of Regent House, Britannia Walk, London N.1, in collaboration with the Geographical Association, is bringing out packs of stereoscopic air photographs with appropriate equipment for their interpretation including notes and exercises. These should be very useful for this kind of project.

4   Preparation of the Observation Exercises to be attempted during the Flight

In the example given in Figure 14, a base map from the 1″ Ordnance Survey map, Sheet 121, was prepared showing the major pattern of drainage and communications and the important landmarks. A line of flight from Castle Donington Airport to the southern limits of Leicester was chosen following the line of the A6 road, and in part the River Soar. Those areas to be given close scrutiny during the flight were termed 'orbital areas', and over them the aircraft was to make a figure-of-eight turn. These orbital areas were indicated on the base map in Figures 14 and 15.

In the aeroplane the pupils may be seated in pairs alongside the windows. One child can be made responsible for ticking off from the flight-list (prepared as shown in Section 5) the features of the landscape as they are observed. The other child can take photographs of these features when the opportunity presents itself. The basic difficulty here, however, is associating the flight exercise with the seating positions of the pairs of children on either side of the aeroplane. Obviously, the pair on the starboard side will have a different view to those occupying the port side, and it follows from this fact that the various features are seen at different times during the flight according to which side of the aeroplane the viewer is seated. This problem can be overcome by providing two sets of flight exercises listing the features in the order in which they will be seen according to the side of the aircraft the observer is sitting. Thus, in the example given here, for those pupils occupying starboard seats the seventeen features to be viewed were lettered from A to Q and those on the port side from 1 to 17. In both sets of exercises the four orbital areas are shown as they occur during the flight. Consequently, on the starboard side, the first orbit, over Mountsorrel, comes into view after the letter H. That is the eighth feature to be noted on this side, whereas from the port side the same orbit is seen after feature number 5.

As the base map was prepared on the scale of 1″ to the mile, it was possible for the flight route to be shown quite clearly on three quarto pages. Each feature to be observed was enclosed in a small circle drawn on the map and marked with the relevant number and letter. The example of the flight exercise given here was prepared for a group of fourth year leavers from a secondary school. It must be emphasised that it is very important, as part of the pre-flight preparation, for the pupils to make a close study of the flight exercise sheets and the base map so that they are quite familiar with what they have to do once the aircraft is in flight.

5 The Flight Exercise

Castle Donington Airport to Wigston, Leicester.
Flight Exercise Sheet: Starboard side.

As they are seen and identified, a tick should be placed against the landscape features listed below, and photographed if possible.

A East Midlands Airport, Castle Donington
B Motorway – M1
C Power Station, Castle Donington
D Communications: i. Cloverleaf intersection on the M1
                                ii. Track of old railway
E College Town, Loughborough
F Reservoirs: A Nanpantan B Black Brook C Thornton
G Beacon Hill
H Swithland Hall

ORBIT A: MOUNTSORREL

i. Ribbon settlement	iv. Track of disused railway
ii. War memorial	v. Buddon Wood
iii. Granite quarries	vi. Swithland Reservoir (crossed by main line railway)

Features of the River Soar to be seen during this orbit:

vii. Floodplain	x. Canal locks
viii. Meanders	xi. Confluence of Rothley Brook with River Soar
ix. Ox-bows	

1 Old sewage works: City Farm, Leicester.

ORBIT B. SCRAPTOFT

i. Old Hall and estate converted to modern college of education
ii. Village and church with modern estate development at the edge of the city

ORBIT C. WIGSTON

i. Factory sites	iii. Shopping centre
ii. Ribbon development of terrace housing	iv. Parks and open spaces

J Straight line of Fosse Way (A46)
K New housing estate, Birstall
L Gravel quarries, Wanlip
M Rearsby airfield
N River confluence: the Soar and the Wreake

ORBIT D. BRADGATE PARK, CHARNWOOD

    i. Bradgate stream

    ii. Cropston Reservoir } Catchment area for city water supply

    iii. Waterworks

    iv. Parkland and woods } Recreational area

    v. Granite outcrops

    vi. Lady Jane Grey's house: Bradgate House

    vii. Old John Tower and War Memorial

O. Loughborough:

    i. University College point blocks

    ii. Grand Union Canal

    iii. Great Central and Midland Railway

    iv. River

P. Brush Engineering works, Loughborough.

Q. Power Station (under construction): Ratcliffe-on-Soar.

6 The Flight

The aircraft took off, made a sweeping turn over the airport before getting on to the course following the A6 road to Leicester. The pilot carefully carried out the instructions given him, taking in all the selected features and performing the four orbits at the places required. The full programme of feature identification, photographing and tape recording was attempted, but a particularly bumpy flight created some difficulties   especially for the few with weaker stomachs than the rest. There were strong gusts of wind nearer the ground, but the pilot did not take the aeroplane higher to avoid these since this would have lessened the clear view of the ground obtainable at this lower height.

The comments recorded by the children using the tape recorder showed considerable enthusiasm for the experience gained from the exercise. 'It was great!', and 'I thought it fab!' summed up the general impression. They particularly enjoyed flying over the parts of the city where they lived. However, the greater complexity of the urban landscape seen from above made the identification of many of the features more difficult for the pupils. Nevertheless, they obtained their quota of photographs, and there was no doubt that they had benefited from this out-of-school experience: one which they might well remember for the rest of their lives, and which had re-stimulated their interest in geography.

7 Follow-up to the Flight

Clearly the opportunity for creative writing as a result of the vivid experience of the flight is an excellent one. Most useful of all, geographically, is a detailed study and discussion based on the comparison with the map and the photographs taken during the flight.

8   A Study of an Airport

A visit to an airport is obviously an excursion which could take place as part of this project associated with the flight. Several kinds of enquiry are possible:

(a) The historical geography of an airport. There are a number of abandoned military airfields, some of which have been partly returned to agricultural use. These sites can be mapped and an attempt made to assess their influence on the landscape.

(b) Information can be obtained from local airport authorities to provide interesting map and statistical exercises. These, combined with a visit to an airport, may be used to build up a case study of the airport.

(c) School parties can make a conducted tour of London airport. This is a very impressive excursion to make if possible. Application should be made to: British Airports Authority, Tours Section, Room 1214, Queen's Building, Heathrow Airport – London, Hounslow, Middlesex.

11

Historical Geography
in the Field

The English landscape itself, to those who know how to read it aright, is the richest historical record we possess. But, 'to read it aright' calls for 'a combination of documentary research and of fieldwork, of laborious scrambling on foot wherever the trail may lead'. The harvest is great but the labourers, so far, have been few. Little fieldwork in historical geography has been attempted in schools. One reason for that is the need for reference to documentary sources at some stage can prove difficult for both pupil and teacher. Another reason may be the subtlety of a great deal of the work. A third difficulty is that, so far, little historical geography as such is taught in the classroom.

However, it is clear that an educational revolution has been and is taking place. It has been marked by greater co-operation between museums, record offices and schools; by the realisation that documents and aerial photographs can be used, albeit with care and preparation, by teachers and children; that such documents can be photocopied for class use and, most important of all, by the realisation that one of the best ways of learning and teaching is to identify and investigate problems in the total environment as they meet the eye. What are the humps in that field? Why is there a sudden right-angled bend in that road? Why does that church stand isolated in a field? Handled correctly such questions can be the beginning of a fascinating reconstruction. The landscape in town and country provides such clues. They have been waiting for centuries to be discovered.

Children can find such fieldwork interesting, profitable and exciting. After all, finding lost villages, tracing the line of a Roman road, examining the geography of a battlefield or the disused track of a railway, all add another dimension to the learning situation. They bring home that sense of the past which is still around us. But what exactly are the fieldwork possibilities for children in historical geography?

1 Settlement Studies

There are numerous field activities aimed towards understanding the site, form and growth of villages, towns and cities. A good deal of comparative work on settlement patterns is possible and a useful basic exercise is the comparison of old and new Ordnance Survey maps with the object of bringing them up to date by fieldwork.

Suggestions for fieldwork in urban areas

(a) *Site factors in the field*

i. Investigate critical site factors in detail e.g. follow the edge of a river terrace marking carefully the boundary between this and river alluvium on a base map. Note and examine in the field fords and bridging points.

ii. Relate castle, market, manor, church or cathedral in terms of location, street pattern. How far is the street pattern focused on these elements? Prepare a model-diagram in the field to summarise your findings. Mark distances apart. Are the elements mentioned still in the town centre or has there been a change in the town's 'centre of gravity' since medieval times? (e.g. Leicester.) If so show the direction of movement by a large arrow on the model. In class later, examine 2½" maps of other towns to see how far there is a typical juxtaposition of mediaeval features.

iii. Field examination of the line of the city walls. Base map with supposed line marked. Follow this in the field noting wall remnants, relationship to present street pattern, indicative street names e.g. Sanvey Gate. A wall traverse might be attempted in such a city as York.

(b) *Town Growth*

i. Urban traverse from the old city-centre, which may have Roman and medieval features, to more recent suburbs along a well-chosen route giving an historical cross-section. Plot age of buildings, relief features, etc. Several traverses in different directions could be profitably compared later. Make into a transect chart in class.

ii. Changing city skylines. Obtain old prints, etc. which give a panorama of the city at different times. Photocopy or duplicate a print. Class to go to the same viewpoint and bring the print up to date by superimposing present-day features on it (see Chapter 4). Photographs taken in a series from the same viewpoint will also give a good skyline. Later, a large panoramic skyline can be drawn in class from this and annotated to show town growth as reflected in the details of the skyline. The field exercise can be attempted from several city viewpoints, if time allows, and later compared as a series.

iii. Field checking and amending of literary descriptions. There are some very good descriptions of townscapes by writers such as Leland, Defoe, Young, Cobbett and Dickens (note his excellent descriptions of London, the impact of railway construction on the landscape). The teacher needs to select an appropriate extract, discuss it in class, duplicate it and take it into the field to study on the spot.

iv. Collect and study location of street names to bring out the growth areas of the town, e.g. Cheapside, Cank Street, Churchgate: medieval; Roberts, Buller, Kitchener Roads: Victorian. Compare with street patterns of different periods. (see E. Ekwall, *The Street Names of the City of London*, 1954.) This may be especially important in an area of clearance or re-development where old names may have already disappeared.

An extension of this can be the collection of house names and builders' names which appear frequently on frontages. The former may be especially useful giving an insight into the social environment in the past and present. It is not unlikely that certain groupings become evident which reflect the town's growth structure. In class a dictionary can be compiled of street and house names related to appropriate maps.

v. House-type studies. Teacher selects a row of houses which show historical contrast.

Duplicates front elevation (as in a survey of a local parade of shops p.79). Adds table to be filled in which requires age, materials, style, roofline features, etc. inserting in the field. Estate agents' photographs will be useful if obtainable.

vi. Comparative studies. A street/area then and now: base map from early edition of 25"O.S. Amend by fieldwork. Redevelopment may have produced a completely new street pattern.

The changing social structure of a street: details of age, sex, origin and occupation of householders can be obtained from the enumerators' lists taken for various nineteenth century censuses and usually kept in the Records Office. If the teacher can obtain these details for, say, one side of a street in 1861, then the class can formulate a questionnaire to be distributed to present-day householders there and a direct comparison made later of the findings.

vii. An evaluation of the historical environment: what parts of the city are worthy of preservation? Preparation: identify these areas on O.S. maps in class. Use Planning Reports to discover future plans. Field visit to one area to determine accessibility, congestion; dangers from re-development, etc.; the state or condition of historical monuments in the area, analysis of visitors to these, possibly by means of questionnaire.

In many British cities, villages have been overwhelmed by urban sprawl. Remnants of such rural 'fossils' can be seen, listed and studied in detail. The story of their disappearance can be traced on O.S. maps and clues still exist on the ground. A good exercise in historical geography would be a careful field/map study of an urban village.

(c) *Changing industrial location*
Select a small area of the city centre. Using directories from a particular year in the nineteenth century prepare a map of the area showing locations of one industry, e.g. hosiery in Leicester in 1862. The class may prepare part of this if the teacher supplies duplicated directory lists and street maps. Once the base map has been completed the class can visit the area to check and amend by fieldwork. How far is the nineteenth century pattern intact? Has the industry moved within the area? Has it moved outside? Have the factories changed in appearance? (consult Oliver, J. L., 'Directories and their Use in Geographical Inquiry' *Geography*, Vol. XLIX, 1964. See also Chapter 18).

(d) *Disease in cities*
Health reports, often dating from the middle of the nineteenth century, are available for many British cities. They contain much information about the incidence and spread of diseases such as smallpox, cholera, typhoid and diarrhoea. Sometimes maps are given to show the distribution of disease together with climatic graphs, etc. An interesting investigation for the class is to consider the geographical background to disease (teacher's aids: L. D. Stamp, *The Geography of Life and Death*, Fontana, 1964; G. M. Howe, *National Atlas of Disease Mortality in the United Kingdom*, Nelson, 1963). Then, with a base map made by the teacher from health reports, investigate an area which, in the past, was a notorious black spot. The following should be marked on the base map:

Waterlogged, ill-drained areas; stagnant creeks or canals.
Waste and derelict land. Note development of squatting here.

Disposal of rubbish, industrial waste, etc.

Condition and age of housing. Classify as good, medium, poor.

Sources of air pollution, e.g. factory chimneys, tanneries.

Follow up by analysing findings and evaluating the quality of the environment from a health viewpoint. Compare with other city areas. Consider micro-climate.

Suggestions for fieldwork in rural areas

(a) *Site and growth*

i. Study in relation to soils, geology, relief, ownership of land, etc.

ii. Bringing the map up to date: use an extract from an Enclosure or Tithe map showing the village in, say, 1783 or 1846, and compare field and road pattern, also extent of built-up area with a recent O.S. map, in the field.

iii. Comparison of land-use using the L.U.S. of the 1930's and amending a duplicated copy in the field to show land-use for 1968. In class, study agricultural changes implied.

iv. Field study of village greens, commons and rights of way. Details for such work are given in *Geographical Field studies in the Durham Area*, University of Durham Institute of Education, 1967. See also W. G. Hoskins and L. D. Stamp, *The Common Lands of England and Wales*, 1963.

(b) *Tracing rural industries*

i. Investigate and map sites of mills, especially windmills. Draw layout plans. Check in *Victoria County History* and H. C. Darby. *Domesday geography* (relevant volume) to discover antiquity.

ii. Search for remnants of rural industries such as cheese-making; Collect details of materials, locations, former importance, processes. Tape interviews on the spot.

iii. Collect unwanted agricultural implements, canal tools, etc. as the basis for a school museum on rural industries. Note where obtained, label and classify.

(c) *Building surveys:*

Old farmsteads, cottages and other examples of vernacular architecture are rapidly disappearing. Yet they represent a wide span of agricultural history, illustrating a series of revolutionary changes in economic and social life and in farming technology. It is the duty of teachers to record such buildings and, indeed, it becomes a fascinating task which reveals a great deal about the region.

Basically the fieldwork involves drawing simple layout plans of buildings and yards, pacing or measuring, noting building materials, questioning to find out about past and present function of buildings, sketching and photographing details to illustrate style and method of construction.

The teacher will find a great deal of help in background, sources and recording techniques from M. W. Barley, J. A. Sheppard, and R. W. Brunskill (see Bibliography). An idea of the objectives and finished work can be gained from: Sir C. Fox and Lord Raglan, *Monmouthshire Houses*, Cardiff, National Museum of Wales (1953–54) and I. C. Peate, *The Welsh House*, Brython Press, Liverpool (1946).

A recording technique suggested by R. W. Brunskill has been adapted for school use and is shown in Figure 16. It is suggested that each child should have such a survey

61

FIG. 16 COMPLETED RECORD CARD FOR SURVEYING VERNACULAR BUILDINGS

GREAT HOUSE	(LARGE HOUSE)	SMALL HOUSE	COTTAGE

LOCATION	COUNTY	MAP REFERENCE	
Ashworthy	Devon	SE 796841	

ADDRESS	ASPECT	WALLING MATERIAL	DATE
High Sharpley	NW	Brick	Lintel 1728

WALL	ADMIXTURE	WINDOWS	ROOF STRUCTURE	ROOF MATERIALS	CHIMNEYS	S.F.
1142	--458	4732	4573	5-44-	75-3-	--72

COMMENTS

Isolated farm house
with outhouses in
rear of local stone

Two chimneys
blocked off and
replaced by partial
central heating (oil)
—
Through passage

PHOTO OR SKETCH

SURVEYED BY	DATE	
B.F.W.	3/4/68	LAYOUT PLAN ON REAR OF CARD

S.F. = SPECIAL FEATURES. CARD PARTLY BASED ON BRUNSKILL, R.W. op.cit. p.47
THIS BOOK IS ESSENTIAL FOR ADDITIONAL DETAILS ON CODING SYSTEM
USED AND FOR IDENTIFICATION CHARTS SHOWING CONSTRUCTIONAL
AND ARCHITECTURAL FEATURES. SEE THE RECORDING OF ARCHITECT-
URE AND ITS PUBLICATION, ALSO NOTES ON THE INVESTIGATION OF
SMALLER DOMESTIC BUILDINGS, COUNCIL FOR BRITISH ARCHAEOLOGY.

card for use in the field. This could be prepared and discussed in class. Large cards summarising class findings might be made as part of the follow-up.

The class should study their plans, sketches and photographs to determine regional styles in building. Special study of one farm in terms of detailed siting in relation to fields, etc. would be additionally useful. Farm inventories, wills and glebe terriers, if published or available in the Record Office, provide the teacher with details of farm or cottage layout and possessions at different periods.

Attention can be concentrated on building stones in farms, houses and villages. An example of a field investigation into chalk as a building stone in a village is given in Figure 17. A base map was completed by observation following an agreed key. Support came from notes, sketches and photographs. If several sample surveys were undertaken in a district it would be possible to work out zones of building stones, i.e. how far they extend beyond their point of origin. Often, the building stones used are no longer quarried e.g. Swithland slate from Charnwood Forest, Leicestershire. A profitable follow-up would be to visit and study abandoned quarries as a prelude to investigating the rise and fall of the particular extractive industry.

FIG. 17

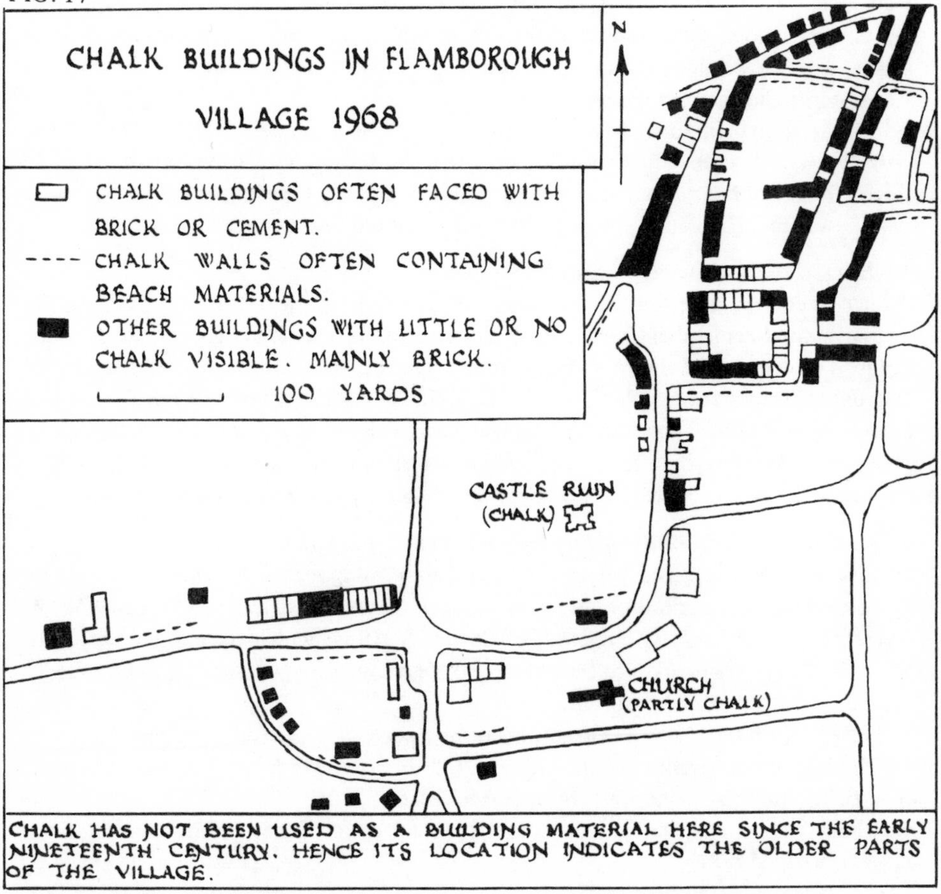

CHALK BUILDINGS IN FLAMBOROUGH VILLAGE 1968

☐ CHALK BUILDINGS OFTEN FACED WITH BRICK OR CEMENT.
---- CHALK WALLS OFTEN CONTAINING BEACH MATERIALS.
■ OTHER BUILDINGS WITH LITTLE OR NO CHALK VISIBLE. MAINLY BRICK.

100 YARDS

CASTLE RUIN (CHALK)

CHURCH (PARTLY CHALK)

CHALK HAS NOT BEEN USED AS A BUILDING MATERIAL HERE SINCE THE EARLY NINETEENTH CENTURY. HENCE ITS LOCATION INDICATES THE OLDER PARTS OF THE VILLAGE.

2 Fields

'Much remains to be done before we shall be able to construe the testimony of our fields and walls and hedges.'

F. W. Maitland.

Scope for practical activities: At first sight an ordinary field may seem very dull with little significance. Closer examination of ground, maps and aerial photographs often reveals secrets which might otherwise remain unsuspected. Some of the most important items for study are: (a) ridge and furrow, strip lynchets, etc; (b) deserted villages; (c) moated homesteads, castle mounds and similar remnants; (d) evidence of mining/industrial activities, e.g. bell pits, salt pans; (e) shape, size and pattern of fields; (f) boundary hedges, walls; (g) field names.

Fieldwork activities

(a) *Recording ridge and furrow:*
Using a 6" or 25" base map for a small area with well defined evidence of ridge and furrow, plot on it by means of field observation the direction, sinuosity, width, edges, headlands, vegetation patterns, standing water, etc. Teacher can prepare duplicated copies of a record sheet similar to Figure 18 which incorporates the base map if he wishes. These record sheets can be issued one between two or three pupils who can then observe and record together.

In class, actual findings may be compared to those suggested by C. S. Orwin, *The Open Fields*, Revised ed., Oxford, 1967. Discussion about relationship of strips to present field boundaries, villages and to physical factors. Compare copy of Enclosure map if available. Film and film-strips on Laxton should also be used.

(b) *How to study deserted villages in the field:*
Though medieval deserted villages may be the main subject for investigation, the teacher should remember that he can find deserted monastic granges, moated homesteads, castle mounds, deserted farms of all periods, abandoned industrial workings, deserted railway stations, old canals in the fields being examined. He might also find himself dealing with 'a town that never was' such as Ravenscar, on the Yorkshire coast, which was laid out in 1920 as a high-class residential and resort town, but never finished and never occupied. Yet its foundations and roads are still visible.

i. *Preparation: a field study of the deserted village of Ingarsby*
In class: what are deserted villages? Why did they disappear? What might you find on the site? Discuss distribution of such villages in your district (teacher uses M. W. Beresford *Lost Villages of England*, Lutterworth, 1954, which contains county lists, pp. 337–393). Mapwork on the siting of Ingarsby using 1" and 2½" O.S. sheets. Teacher obtains permission for visit to site.
ii. *Fieldwork on the site:* no instructions to be given out on arrival. It is important that the children should walk over the site and try to work out roads, house foundations, open fields, mill, fish ponds, etc. by themselves first.
iii. Then, class assembles and discusses findings so far. Together the layout of the village is worked out.

FIG. 18

BASE MAP   LOCATION......................   GRID REF. ....
DATE ....   NAME ....   CLASS ....
CLASSIFICATION: STRIP LYNCHET/ RIDGE & FURROW / OTHER.
SITUATION: GENERAL DESCRIPTION.   APPROXIMATE ACREAGE:

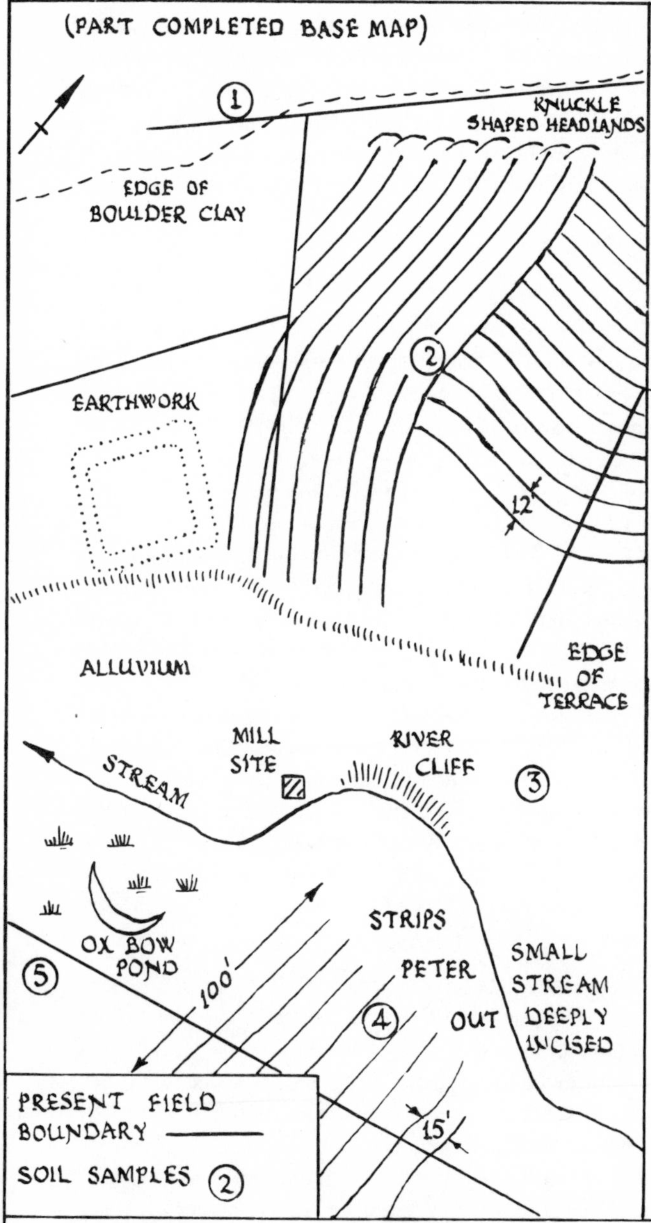

GEOLOGY: FROM MAP & GROUND.
COLLECT SAMPLES.
SOIL: USING AUGER AND SOIL
TESTING KIT MAKE A SOIL TRAV-
ERSE. NOTE UNDERGROUND MASONRY
RELIEF: HEIGHT ABOVE O.D.
MINOR RELIEF FEATURES e.g.
EDGE OF RIVER TERRACE.
LAND USE: WOODLAND, ARABLE,
PASTURE, STANDING WATER.
FIELD ANALYSIS:
i ARE THE STRIPS IN FURLONGS?
ii EVIDENCE OF SETTLEMENT,
  MILLS, ETC?
iii TRACKS, PONDS, ENCLOSURES,
  MOUNDS, MARLPITS?
iv WIDTH OF STRIPS?
v DOES WIDTH VARY?
vi LENGTH OF STRIPS (PACING)?
vii SHAPE IN PLAN: REVERSED S?
viii APPROX. HEIGHT OF STRIPS?
ix DIVIDED BY BANKS/BALKS?
x DO THE STRIPS END IN BUMPS
  OR SMOOTHLY? DESCRIBE.
xi HOW ARE THE STRIPS RELATED
  TO EACH OTHER AND TO ANY
  EARTHWORKS NEARBY?
NOTE FINDS:
  e.g. POTTERY, IRON....
LIST OF PHOTOGRAPHS TAKEN:
SKETCHES:
FURTHER COMMENTS:

(PART COMPLETED BASE MAP)

① KNUCKLE SHAPED HEADLANDS
EDGE OF BOULDER CLAY
②
EARTHWORK
12'
ALLUVIUM
EDGE OF TERRACE
MILL SITE
RIVER CLIFF
③
STREAM
OX BOW POND
⑤
100'
STRIPS PETER OUT
④
SMALL STREAM DEEPLY INCISED
PRESENT FIELD BOUNDARY ———
SOIL SAMPLES ②
15'

THE RECORD CARD IS PARTLY BASED ON H.C.BOWEN'S TALLY CARD IN
ANCIENT FIELDS pp. 63-64

65

iv. Teacher then issues prepared record sheets similar to Figure 19. Maps and record sheets are orientated, then class, either in groups or individually, go over the ground again slowly, filling in the record sheet. Special attention should be given to any remnants above ground, e.g. a ruined church as at Wharram Percy. Soil testing and auger probes should be attempted but under *no* circumstance should any sort of excavation be made.

v. *Follow-up:* findings discussed and collated on return. Large-size class map to be produced from record sheets with sketches, photographs descriptions added. Ground plans and siting of other deserted villages should be compared with Ingarsby (examples given in M. W. Beresford and W. G. Hoskins, *Provincial England.*) Study of medieval new towns and lost ports, e.g. Hedon, Dunwich, can be especially rewarding for those near to such sites as it is often possible to trace wharves, docks, etc. (see M. W. Beresford, *History on the Ground*, Lutterworth, 1957, and *New Towns of the Middle Ages*, Lutterworth, 1967.)

(c) *Fieldwork and field patterns:*
Investigation into field patterns over a small area.

i. *Preparation:* Have the field boundaries changed? Teacher selects part of a relevant tithe map, enclosure map or old edition O.S. after inspection at local Record Office. He may find some useful material already published and therefore more easily accessible. Preferably the fields around a small village should be chosen. Copy and then duplicate relevant part of map. Add a scale.

ii. Class discussion of duplicated map noting well-known landmarks common to past and present, e.g. church, farmhouses, bridge.

iii. *Fieldwork:* Start by viewing fields from church tower to obtain idea of general layout to-day. Divide class into groups each with responsibility for one small section of the map. Each group then compares their old map in terms of field boundaries with the ground for their area. Following an agreed coding system the map is altered to show boundaries which have disappeared, changed or remained the same.

iv. *Follow-up:* Groups then prepare in class one composite map showing changes, if any, in field patterns. This can be enlarged to wall-map size using the epidiascope. Relate to geological and relief background. Note the extent and influence of large landowners. Class discussion on changes: is the field pattern, shape and size related to soil differences? Has fragmentation taken place? Is there evidence of land consolidation, e.g. 'prairie' farming? Find out the effects of social factors, e.g. Gavelkind inheritance in Kent. Have footpaths and roads changed between the date of the map and the present?

(d) *How to study a hedge:*
A combined activity with biologists. Can hedges be dated? Professor Hoskins believes they can be by counting the number of shrub species of plants in a hedge, the oldest hedges having the greatest number of different species. Thus a Saxon hedge might have eight species, a Tudor hedge four, and those of recent times one: a tentative formula of one species for every hundred years of life. Correlation with old maps supports this idea but it still needs a great deal of work before it can be fully verified. However, it does provide a stimulating combined activity which children may be able to test out.

FIG. 19

CAN YOU FIND THE LOST VILLAGE OF INGARSBY?   CLASS ....

LOCATION: EAST LEICESTERSHIRE | MAP REF. .... | NAME .... | DATE ....

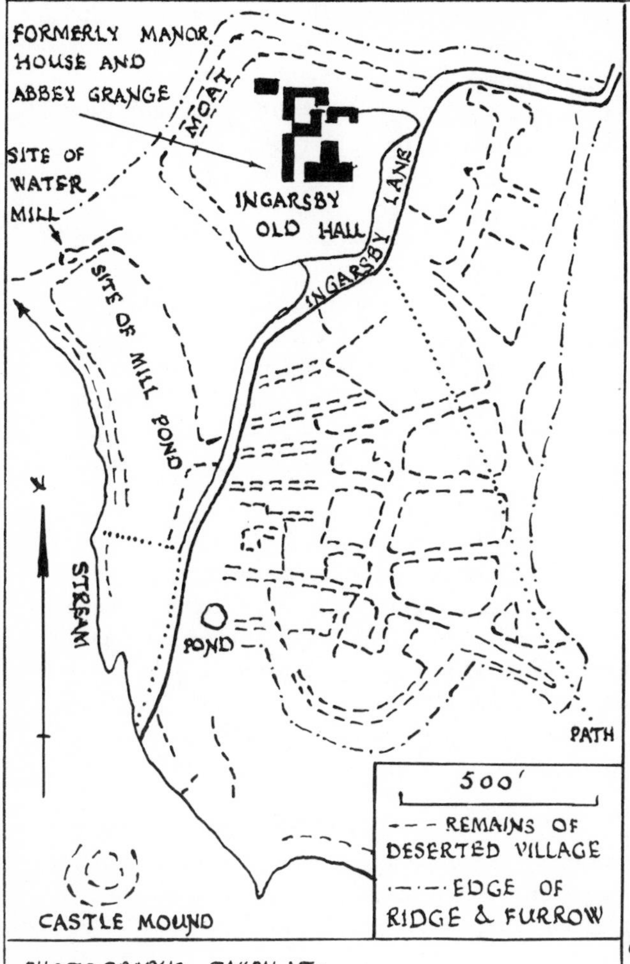

FORMERLY MANOR HOUSE AND ABBEY GRANGE

MOAT

SITE OF WATER MILL

INGARSBY OLD HALL

INGARSBY LANE

SITE OF MILL POND

N

STREAM

POND

PATH

500'

- - - REMAINS OF DESERTED VILLAGE
- · - EDGE OF RIDGE & FURROW

CASTLE MOUND

PHOTOGRAPHS TAKEN AT: ....
SKETCHES AT: ....
FINDS: ....

FOLLOW THESE INSTRUCTIONS:-
1. ORIENTATE YOUR MAP.
2 GO TO THE MAIN PART OF THE VILLAGE. MARK THIS ON THE MAP.
3 WORK OUT WHICH WERE THE MOST IMPORTANT ROADS. MARK HERE i THE WIDTH ....
      ii HOW DEEPLY WORN ....
4 LOOK AT THE MAP. IS THE STREET PATTERN RECTANGULAR, ELONGATED OR ANYHOW? ....
5 IS THE VILLAGE ON A SLOPE OR ON FLAT LAND? ....
6 LOOK AT THE MAP AND REMEMBER HOW YOU CAME DOWN INGARSBY LANE TO GET TO THE VILLAGE:-
   i DO ANY PRESENT ROADS OR PATHS FOLLOW THE OLD ROADS?
   .... ii NOTE ANY SHARP BENDS.....
7 NOW WALK AROUND THE "VILLAGE" ALONG THE EDGE OF THE RIDGE AND FURROW. MARK ON YOUR MAP THE DIRECTION OF THE STRIPS.
8 DOES THE "VILLAGE" EXTEND INTO THIS AREA AT ALL? ....
9 WALK ALONG THE STREAM. NOTE ANY EMBANKMENTS:-
   i WORK OUT THE APPROXIMATE
AREA OF THE MILL POND.......... ii FIND THE SITE OF THE MILL. DESCRIBE IT ....
10 EXAMINE THE POSITION OF THE MANOR HOUSE AND CASTLE MOUND.
11 SUGGEST WHERE THE CHURCH MIGHT HAVE BEEN. ....

IN 1086:   32 PEOPLE IN THE VILLAGE
IN 1469:   VILLAGE BECOMES A SHEEPRANCH. DISAPPEARS.

Stages could be as follows.

i. *Preparation:* Teacher needs to consult old estate maps (16th, 17th centuries), tithe map (19th century), earliest O.S. large-scale maps (see J. B. Harley, 'The Historian's Guide to Ordnance Survey Maps', 1964), and local opinion in order to determine a hedge which seems to have existed for a considerable time. Then he needs to select a short length of about fifty to one hundred yards.

ii. *Fieldwork:* Divide the class into five groups as follows:

Group 1 Flora: how many species? Party has check-list and identification cards (see W. G. Hoskins, *Fieldwork in Local History*, pp. 125–126).

Group 2 Soil: tests to be made on both sides of the hedge. Profiles made at regular intervals to be turned into a soil monolith later.

Group 3 Hedge management and measurement: general condition, details of laying (sketch), height, width, trees in it.

Group 4 Hedge structure and boundary function: note direction and changes in direction; does it cut across ponds, walls, etc. Does it follow a road or track? How is it shown on the O.S. map? (N.B. Word description on 25″ scale). Has it got a ditch or bank? Are any boundary marks visible? Is it a parish boundary? Gateways, stiles, gaps?

Group 5 Photographic unit: photographs to be taken of the entire hedge, then close-up details, e.g. stiles, trees, flowers, shrubs, etc.

iii. *Follow-up:* On return to the classroom a large hedge-chart can be built up by the teacher in discussion with class. A side view of the hedge together with a map of it, illustrated by photographs and field sketches can also be added. An attempt should then be made to evaluate the age and significance of the hedge. Specimens collected should be mounted relevantly to the hedge chart (see Figure 20). A series of hedge studies might be made and compared. After several such efforts certain associations between findings and age will become clear. Map checking should be done by the teacher. It may prove possible to draw a map to show the developing hedge pattern (see Hoskins, *op. cit.*, p. 121).

An alternative to the study of hedges using similar methods might be made in highland areas of Britain (see A. Rainstrick, *Pennine Walls*, Dalesman Publishing Co.). A wall analysis can be made by means of a sketch noting materials, origins, type of construction, general dimensions, boundary function, etc. Since a wall may be difficult to date the teacher would be advised to begin by studying a wall of known or approximate age. The main aim is to discover what typifies a wall of a particular age and thus to be able to recognise and date other walls of that period.

(e) *Practical studies of field and topographical names*

i. *Preparation:* Class to study O.S. maps of the district, say within a small parish, and to collect a list of topographical place names, e.g. the Sike, Holms, Star Carr, Seamer Ings, Seamer Meads, Burton Riggs, Sour Lands, Limekiln Close, Spring Hill, Loftmarishes, Deepdale . . . (from Sheet 93, 1″ O.S.). Teacher and class, using reference books, e.g. Ekwall *Victoria County History*, publications of the Place Name Society, etc., local knowledge and general discussion, work out meaning and possible age.

FIG. 20

A HEDGE STUDY

LOCATION	MAP REF.	HEDGE NO.
GENERAL DESCRIPTION		
MEASUREMENTS		
MORPHOLOGY		
SOIL TESTS		
FLORA CONTENT		
PHOTOGRAPHS		

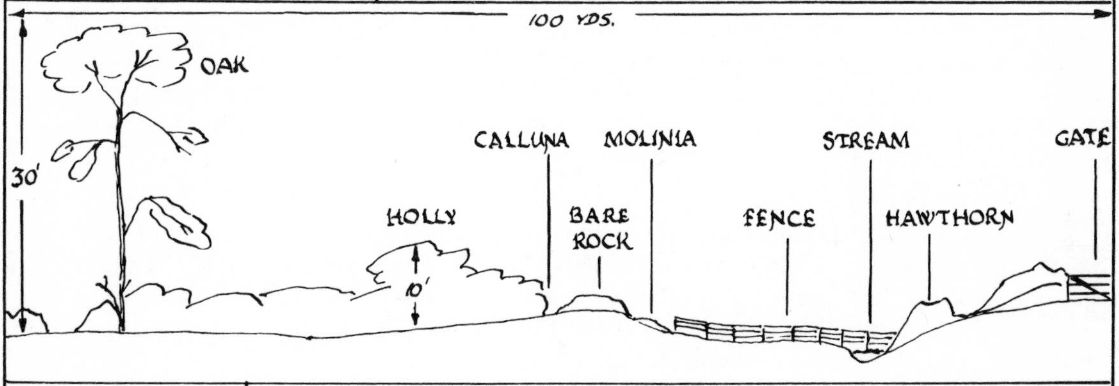

100 YDS.

OAK

CALLUNA   MOLINIA   STREAM   GATE

HOLLY   BARE   FENCE   HAWTHORN
30'   ROCK

10'

pH VALUE	3·8 PODSOL	5·1 BROWN EARTH
FIELD LAYER	BRAMBLES BRACKEN CALLUNA MOLINIA	DOG'S MERCURY
NO. SPECIES	6 2	2

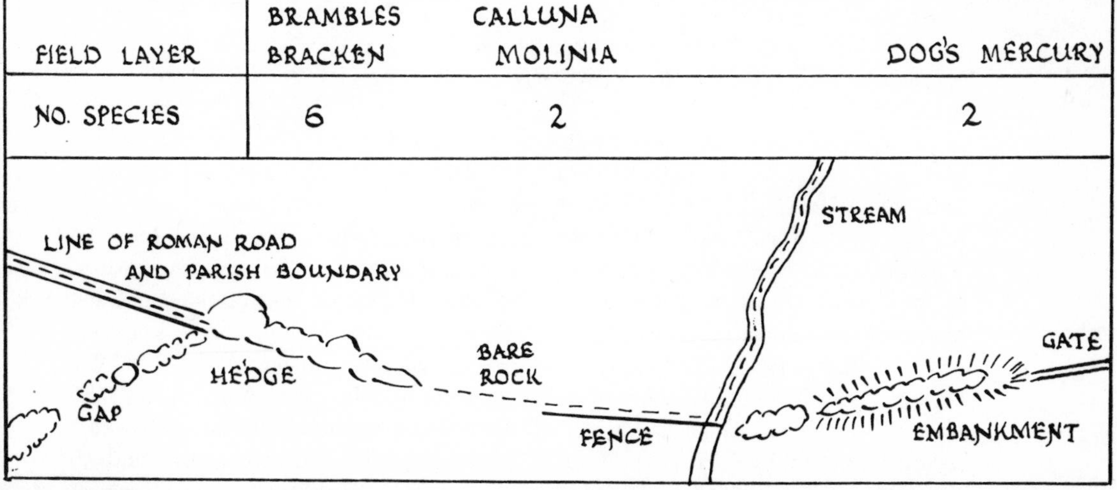

LINE OF ROMAN ROAD
AND PARISH BOUNDARY

STREAM

HEDGE   BARE   GATE
             ROCK

GAP   FENCE   EMBANKMENT

PHOTOGRAPHS, SKETCHES, SOIL MONOLITHS, PLANT SPECIMENS ETC. TO BE ARRANGED AROUND THE CHART.

Individuals prepare a basic check-list of words and meanings in a small pocket-book with simple geology and relief map opposite.

ii. *Fieldwork:* By coach round the parish to evaluate field and place names on the spot, e.g. is there a spring at Spring Hill? Is Loftmarishes still a marshy area or is it drained and cultivated? Insert comments opposite appropriate place name on check-list.

iii. On return class discussion of former landscape and degree of change (teacher's background: H. C. Darby, *Domesday Geography of England* relevant volume. He will then be aware of the extent of woodland, waste, marsh, meadow and cultivated land in the eleventh century). Particular attention should be given to farm and field names, e.g. Sandy Wong, Quebec Farm. If the teacher can discover names from an enclosure or tithe map the party might check with local people to find out if many have been preserved.

3 Linear Earthworks, Hill Forts and Camps

Most British earthworks enclose an area for the purposes of defence and settlement. However, linear earthworks usually consist of a bank and ditch stretching for many miles between two points. They were made for defensive, obstructive or frontier purposes, e.g. Wansdyke, Offa's Dyke (see *Field Archaeology*, Ordnance Survey). Such earthworks, together with Neolithic 'causewayed camps' and Iron Age Hill Forts form a distinctive element in the landscape, especially in the chalklands.

Possible Fieldwork Activities

(a) Map study of the area to be visited noting general distribution and siting of major earthworks especially related to relief, e.g. this comes out well for the Yorkshire Wolds. Next, prepare a tally-card on how to identify an archaeological site with special reference to the site to be visited (see *Field Archaeology*, pp. 8–15). A base map of the site should be duplicated for use in the field.

(b) Annotate base map on the site inserting details showing terracettes arising from soil creep, weathering, angles of repose, measurements relating to the site, e.g. in the case of a hill fort, position of entrance, area enclosed, height of banks, etc. Profiles of banks should be attempted. A full survey would be desirable and might be made in co-operation with the Mathematics Department.

(c) Map reading from the site. Viewfinder exercise. Consideration of the strategy of the siting.

(d) Imaginative reconstruction by means of sketching or description (excellent examples of such reconstructions in Alan Sorrel's work in Methuen Outline Series and Batsford publications).

(e) Co-operation with Biology Department would enhance the possibility of doing the following: i. Soil sampling, with profiles. ii. Study of the effect of rabbits on erosion of the remains. Examination of tree-root systems. Degree of physical, chemical and biological weathering. iii plant frequency and worm count. A belt transect is possible with the former (see J. Sankey, *A Guide to Field Biology*, Longmans, 1958, p. 49). Darwin noticed the importance of worms, especially 'the part which worms have played in the burial of ancient buildings'. Worms can be sorted out by hand from a square 18"x18" dug out to a depth of 12". The total number, multiplied by four, will give worms per square yard. Try to get your pupils to work out how many worms there are to one acre. The result will surprise them. If it is not possible to disturb the site then worms can be obtained from an area of one square yard using a dilute solution of formalin to expel the worms to the surface.

(f) Follow along a linear earthwork as a traverse considering its function as a boundary and/or defence work. How far does it use natural features?

It is clear that the field study of earthworks offers excellent prospects for a joint venture by biologists, historians and geographers.

4   Tracing a Roman road

'After I had studied all the maps I decided to try to get the "feel" of this road by walking along sections which were known and mapped. In this way I almost came to know the Roman surveyor himself . . . .'*Bernard Berry*. There are many Roman roads in Britain, some obvious, others obscure. By following certain clues it becomes a fascinating task tracing and even discovering the course of such roads.

How to trace a Roman road (See diagram over page.)

(a) *Preparation:* Teacher uses O.S. map of Roman Britain and I. D. Margary, *Roman Roads in Britain*, Baker, 1967, to find a short but clear stretch of Roman road nearby, unused if possible, e.g. Wades Causeway (N. E. Yorks), Gartree Road (Leic.). Reconnaissance by teacher. Map study by class followed by preparation of a Clues List. This should include the following:

topographical:
Does road change direction? Any kinks? Straight hedges? Parish or county boundary? Hillside terraces? Any ancient monuments alongside, e.g. tumuli, camps? Bridges or fords nearby? Placenames, e.g. Ridgeway, Causeway, Stretton Magna?

Constructional:
Camber visible? Cutting? Banks, ditches, culverts? Pavements, foundations, maybe ploughed up? Stones, slag, milestones?

(b) *Fieldwork:* Part of class record walk along the Roman road as a traverse. Another group take compass bearings at intervals, keeping a record of changes in direction, noting also distant landmarks. A third group should act as a map reference section carefully following the route on the 2½″ or 6″ map (duplicated copies), noting parish boundaries, etc. All members should have a clues list and report a sighting to main party at once. If the teacher wishes, the course of the road as discovered may be recorded on a master copy of the 6″ map. Otherwise he can design a record sheet which will look like Figure 21 on completion.

(c) *Follow-up:* Large-scale map recording findings to be produced.
Photographs and measurements added where appropriate. Pottery, road metal, etc. to be displayed. Class booklet. Discussion of motorways and Roman roads with, if possible, a later field study of a motorway for comparison.

5   A Castle Study

i. Viewfinder exercises aimed at analysing the strategy of the site.

71

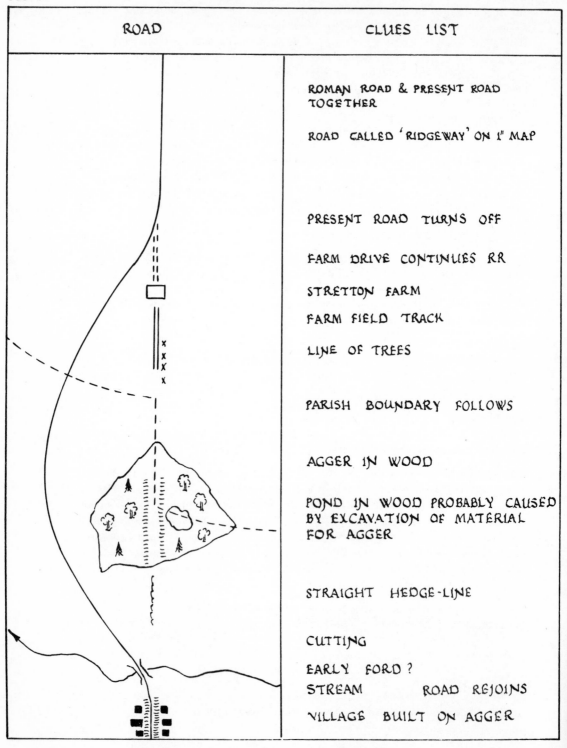

ROAD	CLUES LIST
	ROMAN ROAD & PRESENT ROAD TOGETHER
	ROAD CALLED 'RIDGEWAY' ON 1" MAP
	PRESENT ROAD TURNS OFF
	FARM DRIVE CONTINUES RR
	STRETTON FARM
	FARM FIELD TRACK
	LINE OF TREES
	PARISH BOUNDARY FOLLOWS
	AGGER IN WOOD
	POND IN WOOD PROBABLY CAUSED BY EXCAVATION OF MATERIAL FOR AGGER
	STRAIGHT HEDGE-LINE
	CUTTING
	EARLY FORD ?
	STREAM ROAD REJOINS
	VILLAGE BUILT ON AGGER

ii. Construction of castle in relation to the site, i.e. use of minor relief in defence works, e.g. double fault at the barbican of Scarborough castle.

iii. Base map with castle plan (from Ministry of Works pamphlets). Insert: area of flat land; areas of landslips and erosion; angle of slopes by morphological symbols.

iv. Imaginative reconstruction by sketch or description. If possible compare with a contemporary description.

v. Analysis of building stones, degree of weathering, direction of prevailing winds and effect on weathering. Use hand anemometer. Note wind speed and direction.

vi. Water supply. Wells? Spring? Dew ponds? Aqueducts? Teacher to supply simple geological section.

6   A Church Study

(a) *Site analysis*

i. Viewfinder and map exercises from tower.

ii. Draw simple layout plan. Mark north and note orientation (it may be abnormal, e.g. Rievaulx Abbey).

iii. Site description: relation of church to rest of the settlement. Note isolated churches particularly.

iv. Work on details of site such as marking in the edge of the river terrace on which the church may stand.

(b) *Age and building survey*

i. Establish age of walls by use of 'Observer's Book of Architecture, etc.

ii. Analysis of wall materials. Sketch shapes, sizes, positions. Investigate origins. Note special materials such as Roman bricks.

iii. If possible, do an analysis for several churches in a district so that eventually a map can be produced to show the distribution of building materials.

iv. Make a tower/spire silhouette for as many churches as you can in order to determine local or regional styles (see J. Finberg, *Exploring Villages*, Ch. VIII).

(c) *Cemetery study*

i. On a base map plot location of graveyard, orientation of gravestones, general site details, e.g. on a slope?

ii. Study and sketch tombstones especially where different materials have been used. Later investigate the local industry which produce these. Note inscriptions for details of burials per year, longevity, reasons for death, whole family burials, land-owning families, epidemics.

iii. Any evidence for use as a market? Special features, e.g. the monolith in Rudston churchyard.

(d) Beating the Bounds of the Ecclesiastical Parish.

(e) Examination of old copies of church Visitors' Books; Electoral Roll will give an idea about the catchment area of visitors and congregation. Parish Register will be useful for population structure in any follow-up.

7   An Abbey: an historical geography problem in the landscape

The 'canals' of Rievaulx Abbey in the North Riding of Yorkshire

FIG. 22

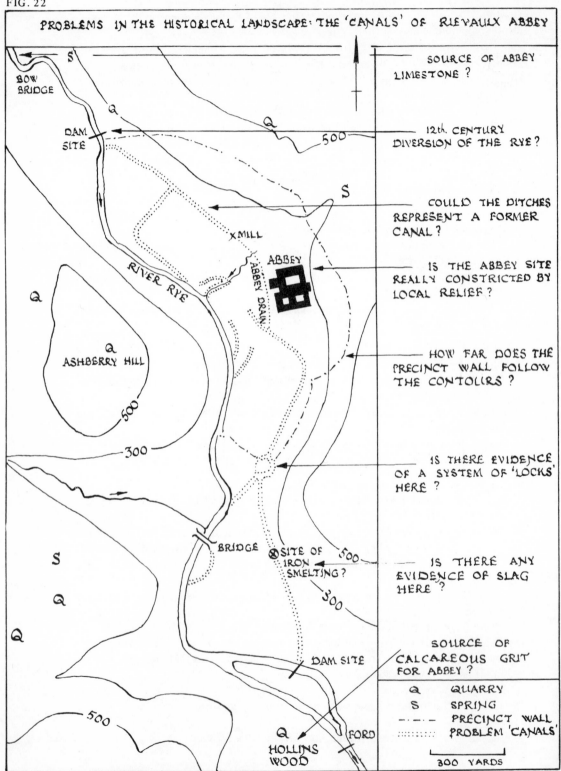

PROBLEMS IN THE HISTORICAL LANDSCAPE: THE 'CANALS' OF RIEVAULX ABBEY

SOURCE OF ABBEY LIMESTONE ?

12th. CENTURY DIVERSION OF THE RYE ?

COULD THE DITCHES REPRESENT A FORMER CANAL ?

IS THE ABBEY SITE REALLY CONSTRICTED BY LOCAL RELIEF ?

HOW FAR DOES THE PRECINCT WALL FOLLOW THE CONTOURS ?

IS THERE EVIDENCE OF A SYSTEM OF 'LOCKS' HERE ?

IS THERE ANY EVIDENCE OF SLAG HERE ?

SOURCE OF CALCAREOUS GRIT FOR ABBEY ?

BOW BRIDGE

DAM SITE

X MILL

ABBEY

ABBEY DRAIN

RIVER RYE

ASHBERRY HILL

500

300

BRIDGE

⊗ SITE OF IRON SMELTING ?

500

300

DAM SITE

500

FORD

Q HOLLINS WOOD

Q   QUARRY
S   SPRING
—·—· PRECINCT WALL
········ PROBLEM 'CANALS'

300 YARDS

A monastic site can provide much interest for the fieldworker. Consideration of the actual siting and layout in relation to physical features; building materials; descriptive writing on the spot; the effect of the monastery on the landscape around are some of the possibilities. Sometimes, as at Rievaulx, a special problem arises. Figure 22 shows the Abbey, the River Rye and some well-marked ditches which are very extensive around the site. What are these ditches? Were they used to transport stone to build the Abbey? Were they used in connection with the iron smelting carried on by the monks? Documents alone do not give an answer.

A field party could spend a very full and interesting day investigating the mystery. They could follow, measure, take soil samples in the 'canals'. They might annotate a base map with their findings. Visit nearby quarries to see if they are, in fact, a possible source for the building stones used. Study the 'canals' in relation to the monastic buildings and look closely at supposed 'locks'. Evidence of early industry could also be examined, e.g. a forge site, mill and slag heaps.

There is, then, unlimited scope for fieldwork in historical geography. It involves all aspects of geography. Space does not allow references to such additional fieldwork possibilities as the geography of Battlefields (for this see *Historical Fieldwork*, edited by T. H. Corfe, in the same series); field enquiries with a tape recorder into local dialects; the study of great parks and landscape gardening; or the growth of the industrial landscape in terms of canal, railway and turnpike studies. But the teacher has a wonderful chance to draw the attention of his pupils to the historical landscape which holds to itself so many secrets.

12

Fieldwork in an
Urban Area

Planning the fieldwork study of a town is not a recondite skill. Many of the same, or similar methods used in more rural settings can be applied in the urban area, although the possibilities for investigation into physical geography are less likely and this aspect of the subject is therefore reduced in scope. On the other hand, it is the historical economic and social factors which are predominant in the town. These may seem unrewarding in their urban complexity for young people to interpret or understand, but careful preparation along the lines suggested here can lead to as stimulating and useful enquiries as any conducted in the countryside. The fact that over 80 per cent of our population lives in an urban environment means, too, that we can correctly combine interest with propinquity!

There is much useful town geography which can be done in the classroom before setting out on fieldwork. This involves using the maps and documents available to analyse the siting of the town with reference to geology, relief and drainage; plotting the distribution of chosen industries using the local commercial and industrial hand-books; or drawing maps to illustrate the growth of the town. The purpose of the fieldwork is to confirm by means of first-hand observation the work accomplished in school and to provide the opportunity of making original maps and diagrams which can supplement or improve upon the published material. Thus, at the beginning of an urban fieldwork study it is necessary to find out from maps the nature of the geology, drainage and relief of the urban area, as well as its boundaries and growth as related to the initial settlement and natural or man-made limitations.

 The teacher must decide from his knowledge of the town which possibilities for fieldwork will be the most profitable exercise and which will stimulate work in the classroom. In the case of schools situated in the larger cities the delimitation of the enquiry might well be set by the limits of the local borough or urban district. However, the fieldwork should always be related to the concept of the town as a whole by using guide books, local histories, models and the town map. The latter should be on permanent display in the geography room for street identification. A record of recent physical changes in the town, such as new housing, the development of point block offices, new roads, etc., can also be plotted and maintained on tracing paper overlays. Articles in the local newspaper often provide information of this kind (and of historical interest too) which are worth filing for reference.

There are three major themes to investigate by means of fieldwork when considering the geography of an urban area. These are: the original site and growth of the town, the effect of communications on the geography of the town; the utilisation of land within the urban area as a reflection of economic and social activities.

FIG. 23

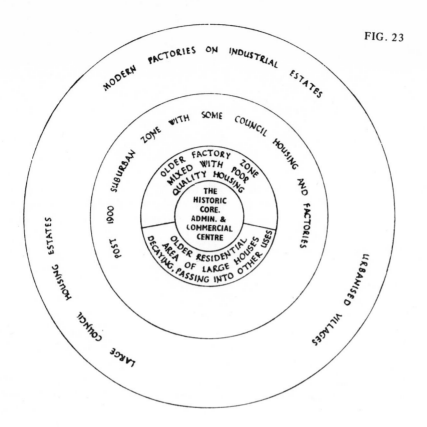

A major concept of value in analysing the problems presented by these themes is that of the urban zone. It is important to understand this concept if the teacher and his older pupils wish to appreciate and fully analyse the geography of their town. In the commonly used terms such as West End, East End, etc., the concept of zone is implicit. The zone, for the purposes of school geography, is that area of a town which has a landscape and land use characteristic peculiar to itself. It is largely homogeneous and can be classified according to one or all three of the following criteria:

1 Functional, i.e., what goes on there – industrial, commercial, residential activities, etc.
2 Historical, i.e., the age of buildings.
3 Social, i.e., categories of people living there.

Applying this concept the student is able to see the differences in spatial patterns and physical environment which occur within the total complexity of the urban environment. It may be better understood by reference to Figure 23 which portrays the simplified schematic zonal development of a characteristic British city, such as Leicester. Using the map of your own town work out a provisional zone map which can be tested by first-hand observation.

1   The Urban Traverse

The traverse method is the best way of making an initial survey of an urban area. This can of course be combined with viewpoint studies (see Chapter 4); and with urban land utilisation mapping. The traverse can be used particularly to demonstrate the siting and development of a town. This means starting the traverse at the historic core and working out across some of the more contrasting zones which also demonstrate the outward urban growth. The traverse route can be given to the pupils as a base map of the roads walked along (25" to 1 mile is the best scale). Box divisions, similar to those in a normal transect diagram (see Figure 12) are then drawn at top and bottom of the base map, and are then used for recording observations made to the left and right of the route. These box divisions can be used for recording urban utilisation symbols, zonal boundaries, industrial and commercial development, buildings of historical or architectural interest, relief and, in appropriate cases, geology, or a suitable selection of these.

For display purposes the traverse diagram can be enlarged to wall chart size and photographs taken by the students, field sketches and, where useful, postcards, can be added to illustrate the record of the investigation.

Traverses can be planned to follow rivers running through town areas, canals or railways. Such investigations would be aimed at illustrating industrial development and urban growth in response to communications. In the case of a river it should also be possible to observe at the same time such features as meanders, river terraces and the flood-plain. In these cases, too, the box divisions are built up alongside a base map of the river or canal in question, or where the investigation is over a short distance the base map may consist of a tracing showing building outlines. The use of the latter, along with changes and alterations, can be recorded directly on to the map, in which case the box divisions are dispensed with.

The major fieldwork activity concerned with communication is, of course, the traffic census. This is dealt with at length in Chapter 13.

2   Urban Land Utilisation Surveying

It is urban land utilisation mapping which can provide a major fieldwork activity. This is because land utilisation is in effect a direct physical expression of historical, social and economic forces which create our urban landscape. Worked at over a period of time, a unique record of a town can be built up which may be valuable not only for its delineation of urban land use, but because patterns of human activity can be observed from it. These may be as simple as realising that estate agents are to be found clustering round railway stations; or as complex as recognising that there have been changes over the years of industrial location within the town concerned.

Methods are given here which can be used with the older primary children or with the more advanced student. They are arranged in a progressive series which aims also to bring pupils to a standard of map reading and geographical awareness so they not only use a detailed classification system for their survey work, but also accurately record and evaluate results. It is also hoped that the student will have a greater understanding of the urban environment: its faults as well as its virtures.

78

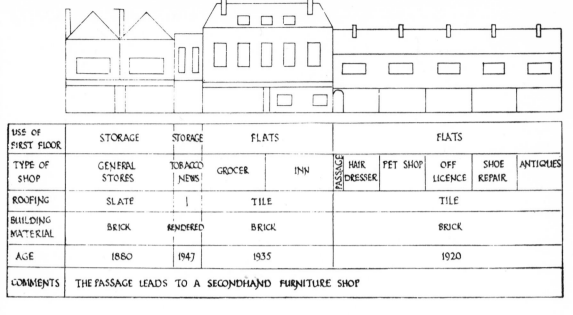

DATE 2·5·64 TITLE: SURVEY OF LOCAL SHOPPING PARADE FIG. 24

USE OF FIRST FLOOR	STORAGE	STORAGE	FLATS					FLATS			
TYPE OF SHOP	GENERAL STORES	TOBACCO NEWS	GROCER	INN	PASSAGE	HAIR DRESSER	PET SHOP	OFF LICENCE	SHOE REPAIR	ANTIQUES
ROOFING	SLATE		TILE					TILE			
BUILDING MATERIAL	BRICK	RENDERED	BRICK					BRICK			
AGE	1880	1947	1935					1920			
COMMENTS	THE PASSAGE LEADS TO A SECONDHAND FURNITURE SHOP										

A Survey of a Local Parade of Shops

Preparation: pupils locate the chosen parade of shops on the map and note its position in relation to the school. The teacher issues boards and paper explaining that a survey is to be carried out and a plan of the shops drawn. An example of what is expected from the children may be drawn on the blackboard.

Fieldwork: visit the parade of shops. Record the number and types of shop and draw a simple elevation and plan of the shops.

Follow-up: plan the layout of the notebook page on the blackboard and discuss any survey problems. Using their survey sheets the class can then produce a fair copy of their work (see Figure 24). In discussion pupils can be asked to divide the shops into a number of different types. They may then draw a graph to show which types of shop are the most numerous. Pupils should be asked to suggest reasons for the results shown by the graph. It may be pointed out that although some shops are much larger than others all are shown as standard units in the graph. This could be used as a basis for another graph based on shop frontages or a survey could be carried out to see which shops are used most by noting how many people enter each shop during a ten-minute period. This kind of work not only encourages careful recording and deduction, but also reveals the limitations of statistics to pupils who will be quick to point out that they might have produced different results if they had carried out their survey on the busiest shop at a different time of day.

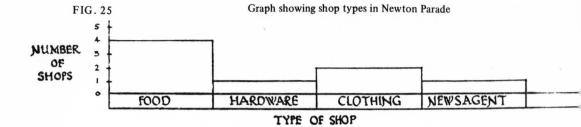

FIG. 25   Graph showing shop types in Newton Parade

The Classification of Accommodation

Preparation: using the local 25″ O.S. map, pupils should locate the school, study the buildings round it, then observe different types of dwelling-place near to the school. Pupils can draw or trace part of the map or the teacher can duplicate copies for the survey. The pupils are asked to suggest various types of accommodation and a list of these is made on the blackboard. Often the meaning of terms such as semi-detached, terrace house or flat are not clearly understood and each term should be clearly defined as the classification A1 to A9 is introduced (see p.86).

Fieldwork: the class and teacher should go out and relate the map to the ground. Plotting of the various types of houses should then be completed.

Follow up: the class draw fair copies of the house classification map. Results can be checked against the map drawn by the teacher on the blackboard. Pupils can plot their results on a graph to show the total numbers of each type of dwelling. Results should be discussed and the implication of recent changes considered, e.g., the increase in the number of flats and maisonettes.

Building Age and Building Materials Survey

Preparation: by reference to well-known local buildings it should be possible to show that the type of materials used for building varies according to the age of the building.

For survey purposes buildings may be divided into seven groups: **I** pre—Tudor, **II** Tudor, **III** Seventeenth Century, **IV** Georgian, **V** Victorian, **VI** 1914–39, **VII** Recent. Pupils can list these groups and make notes on characteristics which will help them to classify building age. They may also be asked to list all building materials used locally. The teacher should check that the list includes such items as: **1** Local stone, **2** Timber, wattle and daub, **3** Slate, **4** Brick, **5** Tile, **6** Weather-boarding, **7** Thatch, **8** Cement, **9** Rendered or pebbledash. In many cases it will be necessary for terms such as weather-boarding and pebbledash to be explained. If each member of the class has lists of building age groups (in Roman numerals) and building materials (in Arabic numerals) they could plot building age and building materials by using the numbers to record their observations on a base map. An alternative method of plotting building materials is to use a colour key on a base map:

80

Wall materials facing the road

BRICK	red
STONE	yellow
WOOD	brown
CONCRETE	blue
PREFAB	green
RENDERED	purple

Plot as a coloured line facing the road.

Roofing

SLATE	blue
TILE	red
LEAD/COPPER	green
THATCH	yellow
TIMBER	brown
CORRUGATED IRON	purple

Plot as a dot on or behind the building.

Fieldwork: these techniques may be used for a class survey or for group work surveys. It is advisable to carry out a class survey first in order to familiarise pupils with the common key and to deal with any survey difficulties.

Follow up: on return to the classroom the results of a survey carried out by pupils using the number technique could be plotted on the colour key basis. The practical advantage of this is that crayons do not have to be used in the field. The completed fair copy maps should be studied in order to determine reasons for any pattern which may appear, e.g., why there are so many houses with slate roofs near the railway station.

The Factory Estate Survey

Preparation: pupils need to familiarise themselves with the colouring used on the new land use survey maps to show industrial land use. They are:

A red wash — manufacturing industry.
Red cross hatching — extractive industry.
Red dot — active tips.
Black dots — abandoned tips or derelict land.
Horizontal red lines — public utilities, including gas, electricity, water and sewage works.

For detailed study of industry, the index number of the fourteen main categories listed in the Industrial Tables volume of the National Census can be used. This is the system used in classifying the manufacturing industries in the new land use survey. The numbers are as follows:

3 Treatment of non-metalliferous mining products other than coal (glass, cement, etc).
4 Chemicals and allied trades.
5 Metal manufacture.
6 Engineering, shipbuilding and electrical goods.
7 Vehicles.
8 Metal goods not elsewhere specified.
9 Precision instruments, jewellery.
10 Textiles.
11 Leather, leather clothing, goods and furs.
12 Clothing.
13 Food, drink and tobacco.
14 Manufacture of wood and cork.
15 Paper and printing.
16 Other manufacturing industries.

A well-organised visit to a factory estate can be most rewarding. As with all such visits the teachers should go over the ground and prepare a base map and question-naire. In modern industries the rate of change and development is so rapid that even the most up-to-date O.S. maps have to be modified to show recent changes. Before or after the teacher's own visit a useful preliminary is to draw up a circular letter and send it to all the firms on the estate explaining the purpose of the enquiry and asking for information on such topics as: date of establishment of the firm, products, raw materials, sources of raw materials, types of power used, sources of power, use of water in processing, source of water supply, disposal of waste, number of workers, whether there are labour supply problems, how goods are transported, how and where goods are marketed, comments on the size and site location of the factory. Requests of this kind usually meet with a generous response from the management of firms. Because they are given some guidance as to what the teacher wants to know, information officers and managers will be more forthcoming than they might be if approached with a brief but general request for 'information'. Such generalised requests yield only a few brochures and sales pamphlets because the firm does not know what is required.

Fieldwork: during the fieldwork pupils should be encouraged to add information in their base map, to complete their questionnaire and to record their own observations. A suitable viewpoint could be used for sketching the industrial landscape and those who feel unable to attempt the sketch could list all large buildings and objects which they can see and write comments against each item in the list. The replies to the circulars sent to firms may enable the teacher to draw attention to significant details which might otherwise be missed or overlooked. Questions should direct pupils to record comments about building layout on the site, site location, access to the estate from main roads, railway siding facilities and layout of factory estate roads. It is often surprising to note on even the most modern factory estates that road layout, access points and parking facilities for both trade and employees' vehicles are either inadequate or badly thought out.

82

Follow up: in addition to making fair copies of their own fieldwork a class display can be mounted by the children. Group reports can be written and duplicated and the firms which supplied the information can be invited to send someone to inspect the display; a copy of the report could also be presented to the firm.

The Grid Survey

Three of the problems met with when a detailed urban survey is carried out are:

(a) preparing base maps which are on a large enough scale and which are both accurate and up to date;
(b) introducing pupils to a method of classification which is both simple enough for them to understand and detailed enough to apply to the complex urban environment;
(c) the recording of the use of those parts of the buildings which are above and below street-level.

The grid survey is a technique which can be adapted to deal with all these problems. Primary school pupils can carry out a survey of this type, but it would probably be unwise to ask them to classify buildings into more than six groups (see Figure 26). For older students divisions under the heading 'Type of Building' can be sub-divided to show different kinds of shops and offices. By increasing the number of columns further categories can be added to the classification list.

In preparing to carry out a survey of this type for the first time the teacher can issue grid sheets to pupils and, using a map drawn on the blackboard, point to buildings, name them and give their use. The pupils can record the information on their grid sheets and the teacher can quickly check to see that they understand how to record information accurately. The value of this 'trial run' will quickly become apparent when the children begin to do their own surveys. During the work of preparation for the survey the pupils should study the local map and locate the area which they are to survey.

Fieldwork: in the first exercise of this type the pupils should survey the same block of buildings. This makes for ease of organisation. Comparison of results will serve as a useful check on accuracy. Subsequently a larger area of the town can be surveyed if the teacher divides the area into survey zones and makes one group of pupils responsible for the surveying of each zone.

Follow up: with this type of survey there is much valuable work which can be done in class after the fieldwork has been completed. Study of the grid quickly reveals interesting distribution patterns. For example, a typical street may reveal that on the corner where the survey commenced there are a number of shops, in the central section there are offices and at the other end of the street there is a petrol station and café. Where possible pupils should be asked to suggest reasons for this kind of distribution. At a later stage when a large area has been surveyed there may be able to divide the area into a number of functional zones. A simple but effective exercise is to ask pupils to shade in red the square on their grid which they think is the most important in their area. They should be prepared to give reasons for their choice. Data

83

FIG. 26

PLACE	PURLEY	DATE	21-3-64	P.8
STREET NAME NORTH STREET	GRID REFERENCE OF STARTING POINT 237524	NAME J. SMITH	CLASS 2A	

POSTAL NUMBERS AND/OR NAME OF SHOP/BUILDING	A ACCOMMODATION OR RESIDENCE	B BUSINESS OR OFFICE	C CATERING	D DEPARTMENTAL STORE OR SHOP	E EDUCATIONAL OR PUBLIC BUILDING	F FACTORY WORKSHOP OR STORE	REMARKS
12	O			/			FLAT OVER GROCER
14	O		/				FLAT OVER CHIP SHOP
16		/					BETTING OFFICE
18							SITE FOR SALE
20		O				/	OFFICE OVER MOTORCYCLE WORKSHOP & SHOWROOM
20	O					/	FLAT OVER SHOWROOM & WORKSHOP

Note. (i) Use a diagonal line to show use of the ground floor of the building and a circle to show use of the first floor.

(ii) The size of shop can be indicated by using each square as a 10-foot frontage unit, e.g. the motor cycle shop has a 20-foot frontage. Alternatively, frontage can be recorded in the remarks column or in a column drawn for this purpose.

from the grid may be transferred to a large scale map of the area; reports and grid sheets submitted by different groups can also be used to build up a composite picture of the area surveyed by the class.

The Classification Sheet Preliminary Survey

Preparation: for effective use of a classification system it is most important that the whole system should appear on one sheet so that the fieldworker does not have to refer to other pieces of paper. The layout of the classification sheet has been specially prepared for use in carrying out a quick preliminary survey. Before going out to apply this classification method pupils should study the sheet and familiarise themselves with its layout.

Fieldwork: the plotting technique is a simple one — put a tick against the appropriate place in the list for each building surveyed. Additional comments can be recorded in the space provided.

Follow up: the advantages of this particular survey are that it introduces pupils to the detailed classification system; it is simple to apply; the survey can be carried out quickly and individual results can readily be checked against those of other members of the class. At a glance the survey sheet provides much information about the character of the area. The teacher may also use the sheets to emphasise the importance of negative evidence. Why, for example, is there no fishmonger or tailor although there are five grocers? In some cases there may be an unusually large number of offices and some of the children may have noticed that most of these are estate agents. This piece of evidence may be used to show that the area surveyed serves a residential area which has a high rate of population change. The weakness of this survey is that it is purely quantitive, there is no record of the location of particular types of buildings.

The Detailed Classification Survey

By using a combination of a letter and a number the use of all buildings in a given area can be recorded. The value of using symbols instead of making a written record will be appreciated by all who have tried to record information in the field. The use of such symbols commends itself to geographers because it enables them to make a graphical record of the information collected. Where possible the letter relates to the use of building: A – accommodation, B – business, C – catering, D – departmental stores and shops, E – educational. The number is used to differentiate between various types of accommodation, shop, etc., e.g., A7 is a flat, A6 is a bungalow. For the complete classification sheet see over. The idea of grouping shops into the five categories, Departmental Stores, Necessities, Common Requirements, Specialities and Services, is one which eliminates a proliferation of unnecessary detail on the map, it also makes pupils think which category an individual shop belongs to when they are making their survey. The examples given under each heading serve to assist them in arriving at a decision. When this type of survey is used for the first time pupils can carry out very useful surveys without using the complete classification system. By using only categories A – E they can get some interesting results. As they become more proficient at the work more categories may be added to their list.

This classification system lends itself to further refinements for the student. To show a change of use of a building the original use can be put in brackets after the classification symbol showing current use, for example, a building which was formerly a large private residence but is now an education establishment could be recorded as E(A1). The use of one letter and one number gives no indication of building age or building materials, but by the addition of third and fourth digits more advanced students could record such information using the numbers listed in the building age and building material survey (see p.81). One of the values of the basic method is that colour is not used, and can therefore be used to emphasise any point of local interest or any special subject of study, e.g., distribution of slate roofing or buildings of the post-war era.

For the survey the teacher should choose a small, clearly defined area. In the first survey of this type it is advisable for all pupils to have a duplicated copy of a large-scale outline map of all the buildings in the survey area. The street plan must be clear

URBAN SURVEY CLASSIFICATION SHEET

A1   Large House in large grounds
A2   Large House in small grounds
A3   Semi-detached or small detached house with garage
A4   Semi-detached or small detached house without garage
A5   Terrace House
A6   Bungalow
A7   Flats
A8   Caravans
A9   Maisonettes

B   Business or Offices excluding banks

C1   Hotels, Inns
C2   Public Houses
C3   Café, Restaurant and Milk Bar

D   Departmental Store

D1   *Necessities*
General
Grocer Provisions
Meat
Fish
Dairy
Greengrocer
Bread, Cakes
Newspaper, Books
Sweets, Tobacco
Draper, Haberdasher
Chemist
Coal

D2   *Common Requirements*
Tailor, Clothing
Jewellery, Watches
Stationery
Hardware
Corn Seed, Dog Food
Wool and Art Needlecraft
Boots, Shoes
Radio, Electrical
Second-hand Shop

D3   *Speciality Goods*
Gloves and Leather
Fancy Goods
Antiques
Furniture
Music
Wines
Millinery
Cars
Gas Board
Electricity Board
Pets
Baby Shop

D4   *Services*
Hair Cutting
Decorating
Shoe Repair
Cycle Repair
Undertaker
Photographer
Monumental Mason
Laundry
Dry Cleaner
Printer

E   Education/Schools

F   Factories

G   Government Post Office
G2   Library, Museum
G3   Police
G4   Fire Station
Labour Exchange

H   Halls, Churches Clubs, Cinemas

J   Residential Schools

K   Lock-up Garage

L   Bank

M   Medical, Doctor, Optician, Dentist, Nurse, Chiropodist

N   Workshop (Electrical Power)

O   Oil and Petrol

P1   Coach Park
P2   Public Car Park
P3   Private Car Park

Q1   Undeveloped
Q2   Public Park
Q3   Allotment
Q4   Cemetery

R   Builder's Yard

S   Sports Ground Playing Field

T   Toilets

U   Utilities Gas/Water Works Sewage Works Electrical Works

V   Special Building (Historical)

W1   Wholesale Warehouse
W2   Storage Warehouse

X   Open Spaces

Y   Beaches

Z   Waste Land Ruined Buildings Old Workings

and a few well-known places marked in so that the map may be 'set' or orientated. Before going out to commence their survey pupils should study the map of the survey area. It often helps if they mark in a few well-known places and street names. Careful briefing on how to plot information is essential because the survey will be of little value unless results are both legible and accurate. Failure at this point is often due to the fact that pupils do not really know what is expected of them. It is best for the teacher to draw a sample field record and fair copy map on the blackboard before taking pupils out to do their survey.

If the teacher prepares a large-scale master map and allocates responsibility for the survey of each block to different groups of pupils (see Figure 27) a composite survey of the whole area may be completed, provided each group produces a fair copy of the map of the area for which they are responsible on the same scale as the teacher's map. As each group complete the map of their block it can be mounted on the master map.

CLASSIFICATION SURVEY ~ FIELD SURVEY RECORD   FIG. 27

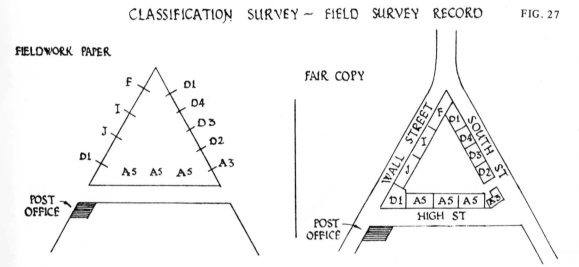

3   Further Suggestions for Fieldwork in an Urban Area

(a) *Roads:* the urban geography of a road. This should be a road of particular interest whose name, course, earlier and present function may have a readily discernible value or is a road typical of a particular zone. For this kind of investigation the questionnaire is useful as shown in the example on pages 88 and 89.

(b) *Trees:* the pupils plot on a 25″ to 1 mile base map the distribution of the larger trees growing in the area. This is a useful exercise emphasising the value of trees in an urban environment and is also a method of teaching tree identification to town children.

(c) *House and Road Patterns:* from type examples of zones up 'specimens' of typical road patterns, architectural styles (period and type); ground plans; dates on buildings, etc. This can be assembled on a 'housing circle' as shown in Figure 28 after the children have been sent out on assignments to collect the relevant material.

87

FIG. 28

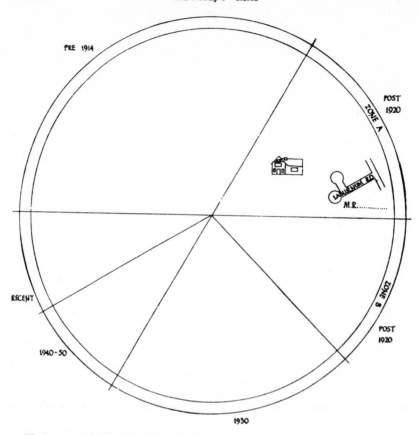

The house can be drawn or photographed.

The road plan is traced from a map of suitable scale.

More than one outer circle can be used for adding extra information.

This diagram is intended to be wall size and is a means of recording the results of class fieldwork in contrasting zones. The sectors are proportional in size to the size of the zone.

Example of Road Investigation Questionnaire
Russell Road. 1880 Lower Income Zone

TIME: 11.40 to 12.15.

AIM:

To examine a typical nineteenth-century road built in response to the increase of the population within the borough after the coming of the railway.

To compare the map with the ground.

To use Urban Utilisation Survey symbols.

METHOD:

1. Look at two old houses in Kingston Road opposite the entrance to

Russell Road. Write in the U.U.S. symbols for them. Photograph or draw them.

What is there significant about the one on the right?

2. Walk down Russell Road. Make notes on the map to show what changes have occurred since its publication.

3. Record some buildings in U.U.S. symbols (on the map). Write down the name of this road and of those turning off it.

4. Consult the geological map. What formation does the south half of the road pass over?

5. Where is there any evidence of this?

6. Note the date on one of the houses, on right-hand side of the road.

7. What would you say the empty building at No. 79A has been?

8. Name the kinds of shops in this road and those near to it.

9. How does this shop grouping differ from that in other zones?

10. Look back down the road from the Broadway end. What building do you see in the far distance? Consult viewfinder.

11. How long is this road?

(d) *Open Spaces:* these include parkland, recreation grounds, etc. Where the school is near enough to one of these the opportunity for practical mapwork is apparent (see Chapter 3); also gulleys and small streams may be used for simpler river work.

(e) *Parish Studies:* the delineation of a parish within an urban area can be a useful exercise. It gives a unit of land which can be investigated along the lines of settlement study already suggested. It also gives an opportunity to enquire into the remains of the rural landscape which may still exist after urbanisation. Such 'fossil' villages are common in any large town or city. Wimbledon is such an example in London. Clues are given here by the roads, pre-urban names and buildings, churches, the origin of open spaces, the course of rivers, age of buildings and so on. Also reference is necessary to the geological map to identify the local geology which may have determined the early site of the settlement. This work can also be connected with investigation into relevant old maps and documents.

(f) *Development Plan Study:* take one aspect of the Town Development Plan. Investigate in any sector or zone of the town the extent to which the plan has been put into operation and mark the changes on a suitable base map.

(g) *Focal Point Study:* on a base map of the home locality children can plot where each member of the class lives and determine the catchment area of the school. Those who deliver newspapers can produce similar maps to show the area served by their newsagent.

Pupils could be encouraged to suggest local focal points which attract people from the surrounding locality, these may be a petrol station, a parade of shops, a bus station and so on. In some cases the sphere of influence of the feature they have chosen can be established from information volunteered by parents, in others it will be necessary for pupils to make enquiries. This kind of practical work can most usefully be undertaken by small groups of keen pupils in their own time with the results being mapped and presented to the whole class. Those who have collected the information can make an oral report on their findings. These can be discussed and each member of the class can then write a report.

4   Synthesis

The above methods for fieldwork in an urban area are not to be thought of as separate isolated exercises. This would be to deny the essence of the geographical approach which attempts to see the environment in its 'wholeness'. In fact the urban study should lead to as composite a picture of the town or urban district as can be produced with the aid of maps, graphs, illustrations and written accounts; also field sketches, photographs (new and old) and completed questionnaires. These all lend themselves to a final display, the value of which has already been adumbrated. If as a sum result of their activities our pupils achieve an understanding of the growth of their home town and a critical awareness towards the problems presented by the enormous changes now taking place in the urban environment, then, surely, our geographical fieldwork has made a valuable contribution to the educative process.

13

Making a Study of Communications

The aim of this chapter is to describe the use of techniques in the study of communications.

Airfields, paths, roads, railways, navigable rivers and canals all form conspicuous geographical features, but their greatest importance lies in the function they perform. The pattern of settlement, agriculture and industry is closely related to the pattern of natural and man-made communications.

The study of communications is often included in fieldwork of a general nature, and it is intended that parts of this study might lend themselves to such a use rather than that the whole should be attempted as a school excursion.

The pattern of communications is still evolving. The opening of a new by-pass will change traffic patterns overnight, with consequent decay in one area of the town and development in another area. In order to make some assessment of the progress of this evolution it is necessary to study the historical development of roads and railways against the background of the local economic and political pressures which shaped the communication system.

As communications reflect the constantly changing needs of society, the fieldworker needs to estimate:

1 The importance of a place as a nodal point.
2 How much the normal traffic pattern and traffic flow vary during 'peak' periods.
3 To what extent public transport is a decaying form of communication.

Attempts should be made to assess the relative importance of, and the inter-relationship between, commercial, private and public transport. Data which is acquired by the fieldworker can be compared with data collected by organisations such as the Ministry of Transport and the British Road Federation. It can also be used in comparative studies of other areas.

Railway traffic does not lend itself to the same forms of study as road traffic. There are many more factors which cannot be ascertained by any amount of observation in the field. The amount of goods traffic on a line may depend on the number of firemen available, rather than on some observable fact. A talk with the local station master and the study of local timetables will assess passenger traffic more accurately than 'train spotting'. In the field the following points are worthy of observation and consideration:

1 Relationship of railways to relief.
2 Methods of engineering and, arising from this question, the extent to which the railways form a physical barrier to other communications.

3 Development, modernisation or decay of railways and railway property.
4 Urban and industrial development bearing direct relationship to the railways.
5 The relative importance of long-and-short-distance trains, goods traffic and passenger traffic.
6 Local factors, e.g., Commuter traffic in suburban areas.
7 Evidence of road-rail co-operation.
8 Methods of motive power and sources of fuel.

A variety of road traffic survey methods are shown. Each has a different aim and may be used to form the basis of a study on its own. They could also be used collectively in order to gain the fullest possible information about a particular area.

Three types of survey, based on the measurements and methods of the Road Research Laboratory and the Ministry of Transport, are selected as suitable for school use. The first two have been tried and tested in their simplest form with nine-year-old children, the third is more suitable for secondary school pupils. These surveys are: (a) Classifying, (b) Directional, (c) Through and Stopping Traffic.

1 Classifying Count

Aims: (a) to classify traffic moving in both directions on one road.
(b) To measure vehicle density.

Introduction: For good results with a large class this survey should be carried out on trunk roads that have wide verges.

Apparatus: Recording sheets on boards (see Figure 29), pencils, watches and measuring tape.

Method: The count is made in Passenger Car Units (P.C.U.): Pedal Bicycle – ½ P.C.U. Light Vehicle to 30 cwt. – 1 P.C.U. Heavy Vehicle over 30 cwt. – 3 P.C.U.

To distinguish whether or not a vehicle should be recorded as above or below 30 cwt., observe the wheels on the rear axle: vehicles with four wheels on the rear axle are counted as being over 30 cwt.
Students may be divided into four groups: Group 1 – counting passenger vehicles travelling towards A. Group 2 – counting goods vehicles travelling towards A. Group 3 – counting passenger vehicles travelling towards B. Group 4 – counting goods vehicles travelling towards B.

Ministry of Transport records are based on a sixteen-hour count in August: children will probably find one hour the useful limit of time for a survey of this type.
Adult enumerators can record 12,000 vehicles per hour unclassified or 800 vehicles per hour if they classify three or four types.

Follow up: (a) Compare results with the records published by the Highways Department of your Local Authority and with the Basic Road Statistics which can be obtained from the British Road Federation, 26 Manchester Square, London W.1.
(b) Compare results with the standards set out in the Ministry of Transport publication Memo. 780. These indicate the traffic loads which are considered

FIG. 29

CLASSIFYING COUNT SHEET

CENSUS TAKEN AT _____   ENUMERATOR _____

DATE _____ TIME : FROM _____ TO _____ DIRECTION _____

	1 2 3 4 5 6 7 8 9 10 11 12 13 14 15 16 17 18 19 20 21 22 23 24 25 26 27 28 29 30
PEDAL CYCLES	
MOPEDS (1)	
SCOOTERS (2)	
SOLO MOTOR-CYCLES	
M/C COMBINATIONS	
CARS AND TAXIS	
BUSES, COACHES & TROLLEY BUSES	
LIGHT VANS (4)	
OTHER COMMERCIAL VEHICLES — 2 AXLES	
OTHER COMMERCIAL VEHICLES — 3 OR MORE (RIGID)	
OTHER COMMERCIAL VEHICLES — ARTICULATED OR TRAILED	

1. Cycles with motors and pedals. | Do not count rollers, barrows, animals ridden
2. Motor cycles with small wheels. | or led, military convoys or unusual traffic.
3. Light three-wheeled cars, invalid carriages, cars with trailers and dual-purpose vehicles used as cars.
4. Three-wheeled goods vehicles, pedestrian-controlled vehicles and dual-purpose vehicles used as vans. All commercial vehicles under thirty hundredweight unladen.
— 5. All commercial vehicles other than (4). Tractors and traction engines.

93

reasonable for various types and widths of road. Comfortable travel is possible if the traffic load is no greater than the free flow capacity of the road (see table):

Width of Road (ft.)	Type of Road	Free flow capacity
24	Two-lane carriageway	6,000 P.C.U.s per 16-hour day
33	Three-lane carriageway	11,000 P.C.U.s per 16-hour day
48+ central verge	Dual two-lane carriageway	25,000 P.C.U.s per 16-hour day
66+ central verge	Dual three-lane carriageway	Over 25,000 P.C.U.s per 16-hour day

Attention should be drawn to the fact that there is no correlation between these classifications and those used by the Ordnance Survey. Use can be made of maps published by oil companies.

2   Directional Count

Aims: to measure and classify the directional flow of traffic through a crossroads.

Apparatus: traffic count sheets (Figure 30), pencils, watches, summary sheets.

Method: five yards from the corners, enumerators stand facing the crossroads, with the traffic on their right travelling from beyond them towards the crossroads. The traffic is recorded in column according to the direction it takes at the crossroads. Children can be divided into groups, one for each corner. Each member of the group should be responsible for making recordings in one particular column. It is advisable to carry out a survey of this type at A or B class road junctions where there are wide verges or pavements.

Follow up: data collected in the field on count sheets should be transferred to summary sheets (see Figure 31). The totals from the four traffic count sheets have to be entered up under the appropriate column of the summary sheet. These totals should be given in P.C.U.s, unclassified except for pedal cycles. The latter are totalled separately as they may constitute a particular hazard at certain times. The pedal cycle may be a special factor if cross-roads are near factories or schools.

To show the relative importance of traffic on the routes surveyed segments of a circle can be coloured in proportion to the traffic volume
e.g., Route A, 400 P.C.U. Route B, 150 P.C.U. Route C, 120 P.C.U. Route D, 50 P.C.U.

Large plans of the cross-roads can be drawn and the survey results can be shown graphically by coloured arrows of proportionate width (see Figure 32 below).

Another arrow may be drawn to indicate the volume of traffic flowing across the junction from the opposite road. The extent of the area where the arrows cross is a measure of the traffic flow difficulties which the Highway Authorities have to deal with in one of the following ways: installing traffic lights, making a roundabout, constructing by-pass, flyover or underpass, regulation of traffic flow, e.g., prohibiting

94

FIG. 30   TRAFFIC COUNT SHEET

NAME OF JUNCTION_____   REFERENCE _____

LIGHT MOTOR VEHICLE : LORRY : BUS : MOTOR-CYCLE : PEDAL CYCLE

LEFT					STRAIGHT					RIGHT				
L.M.V.	LO.	BUS	M/C	P/C	L.M.V.	LO.	BUS	M/C	P/C	L.M.V.	LO.	BUS	M/C	P/C

TRAFFIC MOVING _____ FROM _____
                   (N.-S.-E.-W.)   (NAME OF ROAD)

DATE _____   COMMENCED _____   ENDED _____

CENSUS TAKEN BY _____

FIG. 31

CENSUS OF TRAFFIC AT ROAD JUNCTION

SUMMARY SHEET ~ VEHICLES

NAME OF JUNCTION_____ DATE_____

| TIME | ENTERING JUNCTION FROM | | | | | | | | | | | | PEDAL CYCLES | | | |
| | A | | | B | | | C | | | D | | | | | | |
	L	S	R	L	S	R	L	S	R	L	S	R	A	B	C	D

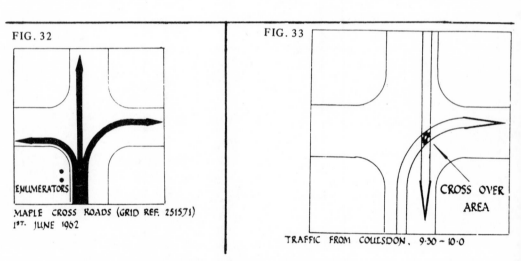

FIG. 32

MAPLE CROSS ROADS (GRID REF. 251571)
1ST. JUNE 1962

FIG. 33

CROSS OVER AREA

TRAFFIC FROM COULSDON. 9·30 ~ 10·0

right turns. These points and the subsequent effect of any necessary engineering may be worthy of discussion. (See Figure 33.)

3   Through and Stopping Traffic Survey

Aims: (a) to study the pattern of the flow of traffic. (b) To give experience in statistical methods.

Introduction: The collection of statistics is essential for sound planning and work of this type has important social and civic aspects. This survey helps students to study the pattern of through and terminating traffic and it reveals the routes used by vehicles

96

within the survey area. Small groups of pupils at secondary schools, who are over the age of fifteen, or students at technical colleges could use this kind of survey; a geography club or society might find it a rewarding morning's work. One survey carried out over a period of half an hour will show which are the busiest routes and it will give some indication of the proportion of through and stopping traffic. If such surveys are carried out at different times of the day or on different days of the week they will provide a useful basis for comparative studies. Data may have to be collected over a period of between two and five hours before it can be used to suggest the purposes for which vehicles entered the survey area, but the sifting of data collected over such a long period becomes and arduous task and after half an hour the plotting technique becomes repetitive.

Method: there are several varieties of this type of survey which as a group are called Origin and Destination Surveys. The method most suitable for use by school children is known as the Registration Number Survey. Enumerators work in groups of three at observation points which have been selected so as to obtain a clear view of oncoming traffic without obstruction to pedestrians. At each check-point two groups of enumerators will be on duty, one group checking incoming traffic, one group checking outgoing traffic. The maximum recording rate per group is five hundred vehicles an hour. Each group of enumerators should possess a watch and score sheet. Before leaving the briefing assembly, watches should be synchronised. Enumerators then take up their allocated positions at the checkpoints shown on the map. At zero hour one enumerator commences calling out the time in minutes and the second enumerator calls out the numbers of passing cars (not the registration letters). Thus: 'Minutes 00 – 776, 32, 479, Minute 01 – 348, 697, 421, 648'. The third enumerator is responsible for recording the information on the score sheet. If there are four enumerators in each team one of them can record the P.C.U. value of the traffic. Other personnel could check the parking load within the cordon area. (See Figure 34.)

Follow up:
(a) On return to the classroom the teacher can ask each group of enumerators to give the total number of vehicles which passed their checkpoint. This information can be summarised on the blackboard, e.g.:

	Incoming	*Outgoing*
Checkpoint A	291	372
Checkpoint B	101	92
Checkpoint C	380	225
Checkpoint D	50	88
Grand totals	822	777

These figures show which are the busiest roads and the difference between the grand totals gives some indication of the proportion of through and stopping traffic.
(b) Teachers may ask individual groups who recorded incoming traffic to give the registration numbers of cars which passed their checkpoint in the first minute. These

FIG. 34

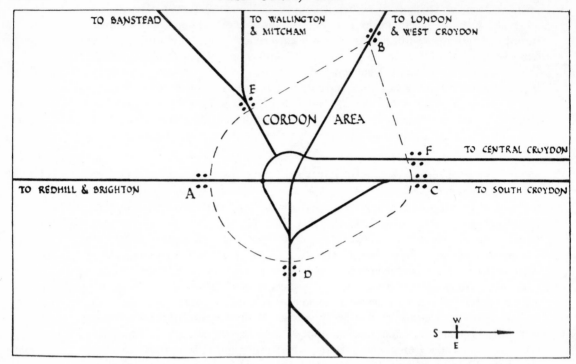

PURLEY CORDON AREA

can be listed on the blackboard and enumerators who were at outgoing checkpoints can look through their lists to try and find the registration numbers. By sampling in this way, the busiest routes are quickly identified and the normal journey time can be calculated.

(c) To complete the follow up, each team of enumerators fills the first two columns on two score sheets. They keep one copy and pass the other one to each of the other groups.

THROUGH AND STOPPING SCORE SHEET

Check Point A INCOMING traffic, Zero Hour 9.30 a.m.

Vehicle No.	Recorded Entry Time (REN)	Route of Exit	Recorded Exit Time (REX)	Duration of Time in Cordon (DUC)
694	00	S	No Record	No Record
372	—	P	04	4
496	—	S	12	12
421	01	D	04	3
		etc.		

SCORE SHEET SUMMARY

	Minutes	Total of Vehicles		
Delay Group I		5	Vehicles – No Recorded Exit	12
Delay Group II		10	Vehicles – No Recorded Entry	4
Delay Group III		15		
etc.				

The score sheet shows that car 372 entered at A, took 4 minutes on its journey through the cordon area and left at P. The number of minutes spent within the cordon area (DUC) indicates the delay group, e.g., if a vehicle's DUC is 3 it is recorded in Delay Group III on the score sheet summary.

As enumerators receive score sheets from their associates they are able to fill in the third and fourth columns of their own score sheet. They will quickly realise that there is an average time taken by vehicles which pass through the cordon area without stopping. If this is five minutes they will start looking for the vehicles which they are trying to trace on the enumerators' list at a time five minutes after the recorded entry time. When columns three and four have been filled enumerators have enough data to enable them to work out the DUC and fill in the other column of the score sheet; they can then complete the score sheet summary.

(d) From the information on the score sheet each group of enumerators can draw a graph. Each vehicle is represented by a line, the length of which is proportional to the duration of time spent within the cordon. Enumerators first plot all vehicles with no recorded exit time. Other vehicles are then plotted in order to time the period spent within the cordon. All lines representing vehicles going from A to S can be given a common colouring. This can be repeated for vehicles using other routes through the cordon area, each route having its own colour.

If there are more than three enumerators at each checkpoint the type of vehicle could be recorded on another column on the score sheet, and this information can also be shown on the graph. Comparison of the graphs drawn by each group of enumerators shows the similarities and differences of traffic using the various routes surveyed. These graphs also give some indication of the parking load within the cordon.

(e) To build up a complete record the teacher can draw a grid on the blackboard and ask each pair of enumerators to give the number of vehicles they recorded as passing through the cordon area within each of the delay groups.

CLASS SCORE SHEET SUMMARY

Delay Group	1	2	3	4	5	6	7	8	9	10		
	1	3	5	10	11	5	4	7	4	10	Vehicles No	
Enumerators	2	5	10	15	20	20	10	15	10	5	Recorded Exit	23
Totals	1	10	10	5	4	5	3	2	4	5	Vehicles No	
	1		5	4	2	7	3	6	2	2	Recorded Entry	8
	0	3	5	2	3	3	11	3	10	4		
	0		2	7	2		6	2	3	2		
Grand Totals	5	21	37	43	42	40	37	35	33	28		

(f) From the data collected a graph can be drawn. The maximum time that a vehicle takes on a journey through the cordon area is assumed to be a time space beyond the maximum vehicle point on the graph. The maximum journey time is calculated as being the time when the maximum numbers of vehicles are recorded plus the time the traffic lights take to turn from red to red. If necessary, the maximum journey time can be accurately established by practical tests. During a period of peak hour traffic, but avoiding times of unusual difficulty when roads are under repair, a test vehicle is used. The driver and observer wait by the roadside at least two hundred yards from the entry checkpoint. Having noted the time and weather conditions the observer tells the driver to follow the next vehicle and notes the time of passing the checkpoint. Should the vehicle being followed turn off the main route the driver of the test vehicle follows the main traffic flow. The observer notes the time as the car passes the exit checkpoint. This procedure can be repeated for each route. The longest journey time is then assumed to be the maximum journey time for through traffic.

FIG. 35

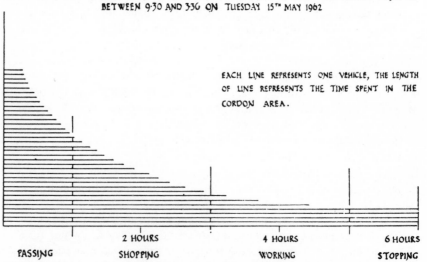

GRAPH SHOWING DURATION OF TIME WITHIN CORDON OF VEHICLES ENTERING AT CHECK POINT A BETWEEN 9·30 AND 3·30 ON TUESDAY 15ᵀᴴ MAY 1962

EACH LINE REPRESENTS ONE VEHICLE, THE LENGTH OF LINE REPRESENTS THE TIME SPENT IN THE CORDON AREA.

2 HOURS   4 HOURS   6 HOURS

PASSING   SHOPPING   WORKING   STOPPING

Graph showing data collected for a Through and Stopping Traffic Survey carried out at Purley on 1.6.62 on all vehicles entering the cordon area between 3.00 p.m. and 3.10 p.m. (see Figure 36).

A represents the duration of survey. B represents the maximum journey time. This will vary depending on the time and area chosen for the survey. In Parley it is about five minutes. All vehicles taking more than five minutes to pass through the cordon area are assumed to have stopped.

(g) The curve of the graph from the origin to point C gives an indication of through traffic, the curve from point C to D may indicate shopping habits; it has been established by research for parking meters that two hours is a reasonable time for shopping. From the statistics can be seen the number of vehicles leaving the cordon area which were not recorded as having entered and this may give an indication of

100

FIG. 36

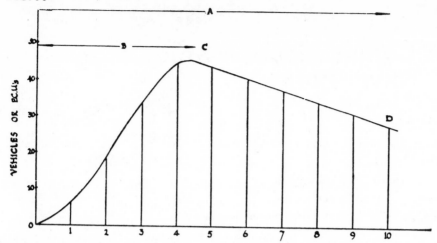

DURATION OF TIME SPENT WITHIN CORDON AREA. RECORDED IN MINUTES

residents leaving the area. The time when these exeats occur may have significance. The number of vehicles entering the area with no record of exit may give an indication of residents' or workers' habits.

(h) If the other origin-and-destination survey methods the only one which can easily be adapted for school use is the interview method. Children could draw up a list of questions in class and interview members of their family asking them to give details of how, where and when they travelled during the previous day.

14

A Communication Study of a Town: Purley

Itinerary

Suitability: work is ranged from adult to nine-year-old levels: selection should be made accordingly.

Time: these sections are each suitable for individual or group treatment; the times for the surveys are given in the notes.
Equipment: O.S. 6″ (Geology Superimposed). O.S. 2½″ TQ 25, 26, 35, 36. O.S. 1″ 170, 171.
Base maps – 2½″ showing roads, railways and contours at 100-foot intervals. Accessibility map of public transport superimposed on base map. Recording sheets (Classifying count sheet, directional count and summary sheets, through and stopping score sheet).

Preparation: study road survey methods of the Ministry of Transport. Obtain permission to walk over British Railways property, and gather relevant information from station master.
    Study local timetables and local transport history.

1   Route

Forester's Drive (300627).
(a) Note Purley's position in relation to Croydon, the Wandle Valley, the Ravensbourne Valley, the Sydenham Hills and the North Downs.
(b) See what traces of Croydon Airport remain. Consider reasons for the closure of this airport, and possible factors to cause reopening or redevelopment.
(c) Consider reasons for the original choice of this site for an aerodrome.
(d) Note the redevelopment of the site in these categories: housing, industry, recreation. How are communications affected by this redevelopment?

2   Route

Walk to Plough Lane.
Note the old route, now terminating at the aerodrome boundary. Consider possible reasons for the original route from church to church, inn to inn or for taking animals to water. (The geological map shows the road route as it was in 1912.)

3   Route

Continue to the Edgehill Road junction with Purley Way (Beggar's Bush).
(a) Note the width of road, choice of route and development of industry.
(b) Purley Way provides an excellent place for a classifying count.

4 Route

Proceed to Kingsdown Avenue (319626).
Walk round the northern edge of the flats to the view point on the lawn in the front of the block.
(a) Note the main valley and tributary valleys and the use made of them by communications.
(b) On the maps identify roads running on ridges and roads running along valleys. Consider which methods of urban railway construction have been used: surface, elevated, shallow tunnel (cut and cover) or deep tube. Find whether the railways have conformed to the roads or the roads to the railways.

5 Routes

Walk to the Royal Oak Centre at the junction of Brighton Road and Riddlesdown Road (320623). Enter the pedestrian precinct of the shopping centre, then go up the stairs to the roof car park.
(a) Draw a sketch map showing the routing of traffic around the centre.
(b) Study the flow of traffic around and through the centre. Note the points at which there is the greatest risk of accident, discuss alternative routing. Plan a traffic count for another time.
(c) Find evidence for previous forms of communication. The tram depot is now a commercial vehicle showroom. You can walk up to Rectory Field to see a section of the Surrey Iron Railway.
(d) The commercial success of this centre depends on the accessibility of the centre to shoppers and office workers. Discuss this in relation to your study of communications around the centre.

6 Route

Continue to the junction of Sanderstead Hill Road with Purley Downs Road. Walk along the footpath by the side of Sanderstead Hill to the junction with Rectory Park and Limpsfield Road (340614).
(a) Note the exposures of chalk, clay with flint and Thanet Sands in the road cutting.
(b) The junction at the top of the hill provides an excellent place for a directional count. Long-distance routes into this junction should be studied on 1″ O.S. maps. Any pupils not required for the directional count could study the old buildings and carry out an archaeological 'dig' near the pond.

7 Route

Continue to Riddlesdown Roman Road (328603).
(a) Note how the road keeps clear of the capping of clay with flint.
(b) Consider other road routes, railway routes and the extension of industry.
(c) This is a good place for field sketching.
(d) The valley can provide a great deal of work bearing on the pattern of settlement and industry in relation to communications. The fact that between Purley and

Caterham there is only ONE shop in the valley south of the railway is the interesting point that would emerge.

8 Route

Proceed to Coombe Wood Hill (321608).
(a) Note the ridge and the flats of Kingsdown Avenue.
(b) Consider the use made by the railways of this valley. A discussion on the reasons for a tunnel at this point should not be started without some knowledge of local railway history.
(c) Field sketches drawn at this point would show many of the physical features relevant to this study of Purley.

9 Route

Walk to Purley Station and enter the goods yard by the Godstone Road entrance.
(a) Note the Quarry face. Observe the varying uses of the goods yard, particularly the new car park for commuters and light industry. A number of cottages near the quarry are flint faced.
(b) From the footbridge note the central position of the station and the size of the installations. Walk through the engineer's department; note that the engine sheds are now used as offices. Nearby are the transformer substations for the electric trains. Return by the footpath over the Tattenham Corner and Caterham branch lines. For a Through and Stopping Traffic Survey, Purley Station would be a convenient assembly point.

Related Studies

Other aspects of communication which could be studied are i. parking facilities, ii. checks on the time vehicles waste on journeys, iii. working out the best routes to use within the urban area, iv. possible improvements which might alleviate the traffic problem.

One interesting fact which emerges from traffic surveys is that no matter how much the volume of traffic increases at a road junction, the proportion using the various routes remains remarkably constant unless some new factor is introduced such as a police regulation 'No Right Turn'.

Where a school carries out annual surveys comparison might be made with the Ministry of Transport's basic figure of 5 per cent for traffic growth. Where the annual growth exceeds this figure calculations of the traffic in twelve years' time could be compared with carriageway capacity figures (as stated previously). Such comparison would show if it is necessary to improve the roads.

In widening the scope of the study of Purley as a local centre of communications, students could map the area served by local schools, shops and newspapers.

The area which is served by an urban centre is often called the urban field. It may correspond to the civic boundary or, as in the case of Purley, to the topographical catchment area. Accessibility is the chief factor in determining the extent and shape of the urban field (see traffic flow map).

ONE MILE

KEY

The area of the circles is approximately proportional to the passenger seats on trains stopping at the stations between 8 a.m. and 9 p.m. (November 1961), 10,000 seats $\cong$ 1 inch radius, e.g. Purley 12,800 seats = $\sqrt{1\cdot28}$ inches radius. (For clarity Purley Oaks Station has been omitted.) Each black line represents two buses travelling between 8 a.m. and 9 p.m. through Purley Corner; or connecting with buses which have passed through Purley Corner during that time. Each line indicates approximately 100 passenger seats, figures indicate the actual number of buses on each route, and the letter 'T' indicates the bus terminus.

By undertaking a fieldwork study of Purley based on traffic surveys, children could not fail to appreciate its importance as a valley junction settlement and shopping centre for a residential area where many commuters live. They would also discover that its industries are alongside an arterial road which carries heavy seasonal traffic from London to Brighton, Eastbourne and other south coast seaside resorts.

15

An Urban Study of a town: Reigate

Aims: to find reasons for the siting of the settlement and to discover reasons for its growth. To practise new techniques of field recording and familiarise pupils with the techniques used on previous occasions.

Suitability: suitable for second year secondary pupils who have had some practice in fieldwork.

Time: 9.30 a.m. to 3.30 p.m.

Distance: Two and a half miles.

Equipment for Teacher: small lightweight blackboard, chalk, programme sheet, 170 1" O.S. map, TQ 25 and TQ 24 2½" maps, geological hammer, gradient measurer, Geological Map 286.

Equipment for Pupils: hardboard with base map on one side, exercises on the other side, small stiff notebook, plastic bag to cover board, pencil, mackintosh, good walking shoes, sandwiches, drinks, one TQ 25 2½" map to each group leader.

Optional equipment: camera, cinecamera, binoculars, compass.

Preparation by Teacher: study maps. Consider means of transport to and from Reigate. If using a coach draw up a seating plan, prepare map and notes for driver. Choose a route and walk over the ground, perhaps taking a few pupils who will be able to act as group leaders for the class expedition. Adapt the route to avoid serious traffic dangers, cross roads at official crossings where possible. Study reference books in local libraries. Produce a sheet of data collected, to issue for consideration after the visit has taken place. Prepare circular to parents, copies of base map, questionnaires and exercises (Base map 6" to mile for pupils plus larger scale for class teaching aid), questionnaire, e.g., What were the main points you noted at stage I? stage II, etc?

Exercises: viewfinder for use on Reigate Hill.
Large-scale 25" plan for High Street with key for plotting building use. (See Figure 38.)

Preparation by class: in this study it is assumed that pupils can set a map, use a variety of building classification methods, check traffic and recognise common O.S. symbols. Pupils take circular for parental signature and contribution to costs. In some cases pupils can prepare their own base maps and question sheet. All can draw up charts for traffic checks, building data and building use surveys in their field notebook. The O.S. map should be studied for the route from school to Reigate. Base maps should be studied so that all pupils know the route to be followed. They can mark the route on

FIG. 38

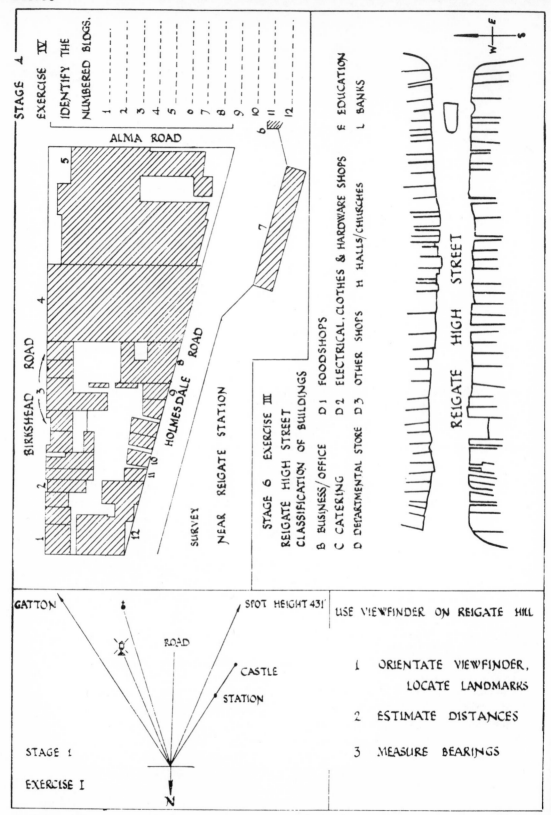

STAGE 4

EXERCISE IV

IDENTIFY THE NUMBERED BLDGS.

1 ....
2 ....
3 ....
4 ....
5 ....
6 ....
7 ....
8 ....
9 ....
10 ....
11 ....
12 ....

ALMA ROAD

BIRKSHEAD ROAD

HOLMESDALE ROAD

SURVEY

NEAR REIGATE STATION

STAGE 6 EXERCISE III

REIGATE HIGH STREET
CLASSIFICATION OF BUILDINGS

B BUSINESS/OFFICE   D1 FOODSHOPS
C CATERING   D2 ELECTRICAL, CLOTHES & HARDWARE SHOPS   E EDUCATION
D DEPARTMENTAL STORE D3 OTHER SHOPS   H HALLS/CHURCHES   L BANKS

REIGATE HIGH STREET

GATTON

SPOT HEIGHT 431'

ROAD

CASTLE

STATION

STAGE 1

EXERCISE I

N

USE VIEWFINDER ON REIGATE HILL

1 ORIENTATE VIEWFINDER, LOCATE LANDMARKS

2 ESTIMATE DISTANCES

3 MEASURE BEARINGS

the map and mount it on one side of their boards. The exercise and question sheet can be mounted on the other side of the board; this avoids flapping papers. Each group leader has an O.S. map and each group has responsibility for a special task, e.g., collecting rock samples or traffic data. The class is told to keep in groups of five under leaders who have instructions as to what to do if they get separated from the main party. With the aid of blackboard examples, the teacher explains new techniques of using a viewpoint and of landscape sketching. Finally, remind pupils of what equipment to bring, arrange when and where pupils will be picked up — thus avoiding any unnecessary journeying to school, and tell pupils what follow-up work they will be expected to do.

1   Coach travel to Reigate

Use 1″ O.S. maps.
Arrange with driver for the coach to remain at Reigate Hill car park until Stage II is completed. Arrange where he will pick up the children — (a) if wet, (b) if dry. The 1″ maps are left on the coach. Children leave the bus and group leaders check that all have pencils, boards and notebooks.

2   Reigate Hill

Route: walk from car park to the best viewpoint 263523.

Note: setting of the town just to the south of the North Downs. Relationship of relief to geology. Mention water supply and draw attention to larger woods of beech trees on the scarp slope of the North Downs.

Activity: discuss reasons for the siting of Reigate in Holmesdale. Set 2½″ map and locate positions. Complete the viewpoint exercise. Teacher draws sample landscape sketch on blackboard. Children study sample, then draw their own sketches. Collect samples of chalk and flint. Use gradient measurer on slopes.

Comment: load display blackboard and gradient measurer on to coach.

3   Northern fringe

Route: down Reigate Hill 263523 to 254507.

Note: type of houses and building materials. Large Victorian houses, showing signs of subsidence on gault clay, are being pulled down for redevelopment of site with high-density housing and flats.
There are examples of infilling.
Study the type and amount of traffic using the N. — S. route on Reigate Hill Road.

Activity: plan drawing of an example of infilling. Traffic census. Fill in building material chart. Elevation study — drawing of sample houses. Collect sample of gault clay.

4   Station area

Route: walk round block by the station 255507.

Note: buildings of late Victorian era.

Shopping area near station. Site of station in relation to town centre. Electric train service terminates here. Railway uses Holmesdale route. Congestion is caused by the level crossing.

Activity: fill in location map. Draw plan of rail-track layout. Discover where trains go to and come from. How many trains (a) go through? (b) terminate here? How many season ticket holders are there? Draw an elevation sketch of Victorian buildings.

5 Castle area

Route: southward from station to castle grounds 253504.

Note: how roads have to avoid or tunnel under castle ridge — evidently the early route was more E. — W. than N. — S. The south side exit gives a moderate view of the High Street area.

Activity: the castle grounds are a convenient stopping-place for lunch. Toilets are near by in the old Town Hall.

6 Town centre

Route: walk along north side of High Street from old Town Hall 253503 to road junction 250503. Cross road and return on south side.

Note: remaining old buildings — old Town Hall, Red Cross Inn, sixteenth-century cottages, nineteenth-century chapel in centre of main street on the southern side. Evidence of traffic problem.

Activity: use a base map to make a record of buildings classified according to age, building materials, use, etc.
Find out the total number of shops which can be classified under the headings:
Food and Necessities, Common Requirements, Specialists, Services.
Estimate the approximate frontage of each shop by pacing. From the information collected work out the total frontage of the shops in each of the classification groups indicated above.
Traffic census. Elevation drawings. If there is a hole in the ground collect rock sample.

Comment: early closing day is Wednesday — a convenient day for this work. There are fewer people and pupils are less likely to stray. A key of letters and/or numbers may be devised to show building use (see Figure 38) on p.108.

7 South of the High Street

Route: go south from cross-roads 253503 to bus station 257498, then into Priory Park to viewpoint 252494.

Note: Eighteenth-century Priory House near town centre — now the Park is a public open space. Study of bus timetables shows how important Reigate is as a route centre (nodal point).

Activity: at the bus station note destination of buses. (Collect rock sample in Park.) At viewpoint 252494 draw landscape sketch to illustrate setting.

Comment: this is a convenient place to summarise field observations and encourage pupils to write notes about their impressions while these are fresh in their minds.

Sum up: importance of road and rail transport in town development.

Not an old town like Guildford or Leatherhead. Parish church is away from town centre.
Ask children what they consider to be the main differences between (a) area north of station, (b) station area, (c) town centre.
Field conclusions: A shopping and service centre for surrounding community of commuters' families.

Comment: for a representative picture of the whole Redhill-Reigate area note differences — Redhill rail junction, different type of shopping centre, factory areas. Coach can wait to pick up children on road at 253492, at the Bus Station, or at some other convenient place where the driver has been instructed to meet the fieldwork party.

Follow up: (a) make fair copies of base map exercises and notebook data for display.
(b) Discuss facts observed, group leaders reporting on site, geological samples, transport, building materials, type of shops in High Street, comparison of elevation drawings of different areas. Remind children of limited value of brief traffic checks.
(c) Fieldwork requires a follow-up study of reference books. To save time a collection of useful reference data can be issued by the teacher.
(d) From field observation, maps and reference data, draw conclusions with emphasis on geographical factors: water, shelter, security (castle), routes (the old high downs road through Gatton and Chipstead, the A 217 Reigate Hill Road, the east-west Holmesdale route and the A 242 Merstham Gap Road), distance from London (electric train service reduces travel time).
(e) Children write up their own account.
(f) Any photos, colour slides or cinecamera shots are useful stimulants to discussion in follow-up work. Often a convenient final summary is provided by a town's crest if the heraldic symbols can be understood.

DISPLAY			
Base Maps	Zones Map	Relief Sections	Geological Map
Transport	Elevation Studies	Growth Graph (From census statistics)	Rock and Plant Samples

Roses
Road
Rail
Traffic Census
Neat copies of exercises — view point, building materials, building location map.
Examples of pacing, classifying and grading methods of building use survey.
Written Accounts. Models. Drawings. Photos. Post Cards.

16
A Parish Study

The Parish of Ewell in the Borough of Epsom and Ewell

1 Introduction

In planning and carrying out a parish study there is a danger of losing sight of the fact that it is the parish as an entity which is the significant feature of the study. The parish boundary is often regarded in local studies merely as a convenient device for limiting the area of study. In this sense, the parish boundary has no more significance than the margins of a map sheet. A series of pieces of fieldwork carried out within a parish boundary do not add up to a parish study unless they are consciously related in some way to the significance of the parish area.

In this sample piece of fieldwork the term 'parish study' has been interpreted in this strict sense. Thus it should be recognised that the programme of fieldwork outlined is, of necessity, very selective and by no means exhausts all the possibilities for fieldwork within the Ewell area. The parish study, however, can provide an excellent purposeful introduction to further field studies within the area but which are not identified with the significance of the parish.

A parish study provides a good opportunity for individual work and private investigation on the part of pupils. Much of this can be done in their own time if they are familiar with some of the techniques suggested in this book. A parish could be studied by a beating of the bounds or by separate studies of the main sub-divisions. In the case of Ewell the latter would involve three major pieces of fieldwork. It is essential, however, that pupils should end with a balanced perspective of the whole unit. With a class of children, a traverse proved to be the best practical way of initially studying the 'strip' Parish of Ewell.

2 Aims

The chief aim is to discover a reasoned explanation of the size, shape and position of the parish. A secondary aim arising from this main theme is to demonstrate the present loss of identity of the parish as a significant social unit.

Thus this piece of fieldwork is an exercise in historical geography. An essential part of this approach is to develop discrimination between natural features and human modifications of those features.

3 Equipment

1″ O.S. map sheet 170. 2½″ O.S. map sheet TQ 26 and TQ 25. 6″ O.S. map sheets XIII S.W. and XIX N.W. old edition or sheet TQ 26. S.W. – the new edition. 1″ Geological Survey Sheet Reigate. Geological hammer, soil auger and trowel. Specimen containers (polythene bags suggested).

112

4   Preparation

(a) Discuss the historical nature and purpose of parishes in general, emphasising that to a large extent they existed as self-sufficing community regions. The immediate region had to provide for most of the basic needs of the inhabitants because of the inadequacy of transport. (Ewell was approachable only on horseback or on foot two hundred years ago. In the reign of George II, Acts were passed for the construction of roads suitable for wheel traffic. An Act of 1755 authorised construction of the road from Ewell to Cheam, the present Reigate Road and the Epsom Road.)

(b) Establish the course of the Ewell parish boundary from 2½" or 6" maps and draw a sketch map of the parish including main communications, drainage features and a reference grid. Duplicate a base map showing this data and provide each pupil with at least one copy for use in the field.
N.B. This study is confined to the main part of the parish and the detatched portion in the Kingswood area is not included.

(c) After suitable instruction, pupils should prepare a relief section extending from Drift Bridge (231601), through the centre to the northern boundary of the parish.

FIG. 39   GEOLOGICAL SECTION THROUGH THE PARISH OF EWELL

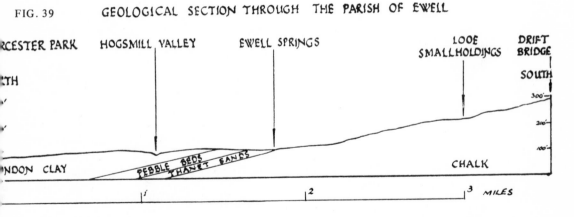

(d) Pupils should be given instruction in methods of recording in the field. Information regarding relief, geology, soils and land use should be plotted on base maps. Detailed information should also be recorded on one of the O.S. maps by a reliable member of the group. Information regarding age and function of buildings should be recorded in note form with grid references.

5   Route

The traverse outlined extends from the southern to the northern boundary of the parish. The starting-point is Epsom Downs Station (225590) and the finishing-point is in Ruxley Lane (206642).

From Epsom Downs Station cross to the open downland. On early maps this area is named Ewell Downs.

113

Note:

(a) Most southerly area of the parish.

(b) The dip-slope northwards to the London Basin.

(c) The open grassland of the Downs.

Activity:

(a) Discover age of station and note building materials used.

(b) Discover average depth of soil on Downs from at least six borings.

(c) Establish that the underlying rock is chalk.

6 Route

Continue northwards along Longdown Lane, noting age and material used in buildings on right-hand side. Turn right into College Road and proceed to the old chalk pit (227602).

Note:

(a) Height of the chalk face.

(b) Dip of the strata.

(c) Present use of the pit.

(d) Land use of surrounding area.

Activity: sketch the face of the chalk pit.

7 Route

Continue to the junction with Reigate Road, turn left and walk to the small holdings at North Looe (230605).

Note:

(a) Shallow dry valley in which the area is sited.

(b) Chief produce.

Activity:

(a) Investigate average soil depth by sample borings.

(b) Establish that the underlying rock is chalk.

8 Route

Continue along Reigate Road to the bridle path to Banstead Road (224615).

Note en route:

(a) Land use of Downs Farm.

114

(b) Age and building materials of North Looe House.

(c) Age and building materials of houses on left-hand side of road.

9   Route

Follow the bridle path to Banstead Road.

Note en route:

(a) Use of adjacent land (playing fields).

(b) The artificial grading of the land.

(c) The site of the Ewell Technical College.

(d) The centre of Ewell Village with its many trees.

Activity: make borings along the track and discover the nature of the underlying rock. (This track crosses a residual area of Thanet Sand. It can be identified by the change in slope.)

10   Route

Continue to the junction with Cheam Road.

Note:

(a) The course of the parish boundary. (At this point it passes through a house!)

(b) Age and building materials of houses in Cheam Road.

Comment The fact that the Parish boundary passes through a modern house shows that, as a demarcation line, it has lost its former significance.

11   Route

Continue to East Ewell Station.

Note: age of the station and building materials used in its construction.

12   Route

Continue to the By-pass, turn left and walk along the By-pass as far as the junction with Reigate Road and return on the opposite side to Cheam Road.

Note en Route:

(a) Horticultural nurseries.

(b) The density of the traffic flow along the By-pass compared with that of the Reigate Road.

Activity: investigate the depth and texture of the soil.

Comment: stress the importance of light loamy soils in intensive horticulture.

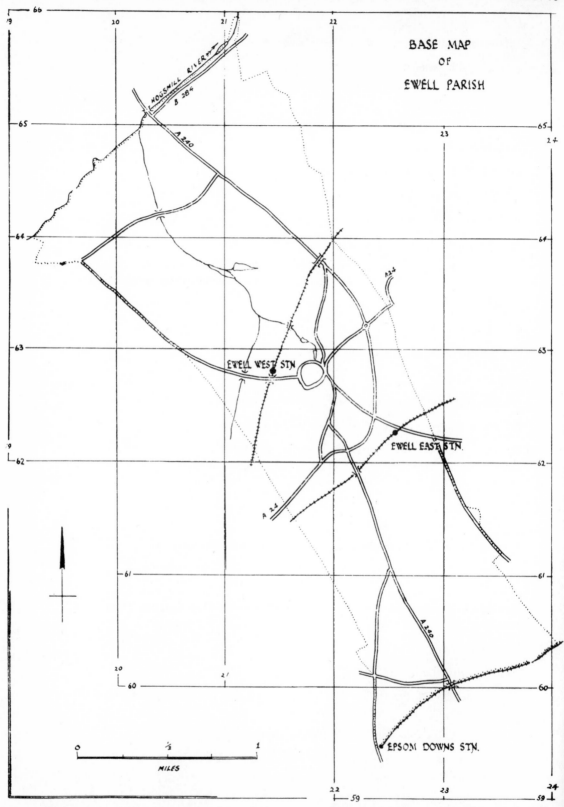

FIG. 40

BASE MAP
OF
EWELL PARISH

EWELL WEST STN

EWELL EAST STN.

EPSOM DOWNS STN.

0   ½   1
MILES

13   Route

Continue to old Marl Pit opposite Pit House. Chalk was quarried here.

Activity: Estimate how long the pit has been disused by noting tree growth.

Comment: Stress the importance of marling land in early times.

14   Route

Continue to the corner of Church Street.

Note en route: Age and building materials of shops, particularly the market stores on the corner of Church Street.

15   Route

Walk up Church Street and enter the grounds of Glyn House. (Prior permission to be sought from the warden.)

Note en route:

(a) The old Lock-up and fifteenth-century flint Church Tower.

(b) Locate the springs adjacent to the main road.

Activity:

(a) Investigate the underlying rock to the east of the Springs (Thanet Sand).

(b) Locate and, when possible, count the actual spring sources.

16   Route

Cross the main road to Bourne Hall Gardens.

Activity:

(a) Trace the outflow of water at the ponds in Bourne Hall Gardens and to Horse Pond by Chessington Road.

(b) Investigate the underlying rock in Bourne Hall Gardens.

Comment: the spring water is very clear — an indication that it contains calcium. The flow of water from these dip-slope springs varies according to the rainfall of six months earlier.

17   Route

Continue along the Kingston Road to Upper Mill.

Activity:

(a) Draw a sketch plan of the mill pond and the mill.

(b) Draw one elevation of the mill noting the materials used in its construction.

(c) Determine underlying rock alongside the river below the mill (Woolwich Pebble Beds).

18   Route

Continue via Lower Mill to the bridge under the railway (216633). Descend alongside the railway embankment to the River Hogsmill and cross by the footbridge. Follow as close to the course of the river as is practicable to the Pack Horse Bridge (210637). The route can be very muddy!

Note:

(a) The river has been artifically straightened.

(b) The banks have been reinforced in places to arrest erosion.

(c) The number and variety of trees.

(d) It is a marshy area and was formerly called Ewell Marsh.

(e) The contrast in grass growth, compared with grass on the chalk downs.

(f) The absence of settlement on the flood plain.

(g) The age of buildings on the margin of the flood plain.

(h) The artificial grading of the flood plain to provide a suitable site for an athletics and recreation ground.

Activity:

(a) Determine the underlying rock (London Clay).

(b) In the river gravel look for fossils (echinoderms).

(c) Draw a sketch section across the valley to show the relationship of the building development to the valley.

(d) Near the Pack Horse Bridge look for evidence that this was the site of a water mill. The main stream used to flow under the Pack Horse Bridge. This course has now been partially filled in. The present stream is part of the former head race. This can be verified by noting the difference in levels between the foot-bridge over the stream and the Pack Horse Bridge.

(e) Draw a sketch of the Pack Horse Bridge.

Comment: the mill supplied gunpowder in the Franco-Prussian War of 1870. Alders and willows grown near by were used for charcoal. Stress the importance of the Pack Horse Bridge as evidence of the state of communications in earlier times.

19   Route

Follow course of river to where it crosses Ruxley Lane. Bus routes go to Ewell, Epsom and Epsom Downs, and to Kingston.

118

Follow-up Activity in Class

(a) Discuss accumulated evidence of geology. Dip and succession of the sedimentaries. Interpolate geological boundaries on the base map.

(b) Add geological section to relief section.

(c) Indicate on relief section variations in depth of soil cover by means of 'shading'. Indicate also variations in tree growth.

(d) Insert on base map the positions of all buildings which were observed to be of pre-railway era. Note position of village nucleus.

(e) On the basis of the recorded information concerning the soils, vegetation and present land use, discuss and reconstruct the probable land utilisation in former times. The following points should be established:

i. The extreme upper area of the chalk dip-slope was used for open grazing lands, especially for sheep.

ii. The main arable areas were situated on the central and lower areas of the chalk dip-slope. These were the sites of the common fields. (It would be a useful further exercise to discuss which particular areas of the dip-slope would produce the best crops, bearing in mind the Looe Smallholdings area and the junction with the tertiaries.)

iii. Main village settlement was on the Thanet Sands and Pebble Beds. The springs are not the only advantage of this site.

iv. The London Clay areas were mainly woodland or marshland. Some of the drier areas would have been cleared to provide cattle pasture. The open commons were in the Ruxley Lane area. A dairy farm, Scott's Farm, was there in 1950.

v. The river provided power resources for milling and a specialised local industry — gunpowder making.

This reconstruction should be recorded by each pupil on a base map. This could then be compared with actual historical records some of which are indicated in the appendix. A large-scale map for class display could be made as a final summary. Children should be encouraged to visit the library to carry out their own research; they should also look for old photographs and picture postcards of the area.

More detailed Follow-up Activities

The theme of the present loss of identity of the parish as a significant social unit could be pursued along the following lines:

(a) Investigation of the reorientation of the main direction of traffic flow through the parish.

In earlier times the main lines of internal movement within the Parish would have been mainly north to south — Reigate Road and Chessington Road — because these roads connect the main parts of the Parish with the village centre. The modern traffic routeways — the railways and Ewell By-pass — trend east to west and bear no relationship to the Parish boundaries. This fact could be demonstrated by a traffic census carried out on the main roads of the parish.

119

(b) Investigation of the places of work of the inhabitants of the parish. A sample **census** could be made in contrasted residential areas of the Parish — West Ewell, **the Village** and East Ewell.

Appendix
POPULATION AND THE CIVIL PARISHES OF EPSOM AND EWELL

	Epsom	*Ewell*
1801	2,404	1,112
1811	2,515	1,135
1821	2,890	1,550
1831	3,231	1,630
1841	3,533	1,622
1851	4,129	1,918
1861	4,890	1,922
1871	6,276	2,214
1881	6,916	3,002
1891	8,417	3,143
1901	10,915	3,338
1911	19,156	3,867
1921	18,804	4,187
1931	27,092	7,117
1937	Borough of Epsom and Ewell	56,700
1951		68,055
1961		71,159

USEFUL REFERENCE MATERIAL

A Short History of Ewell and Nonsuch, by Cloudesley S. Willis. This has a useful large-scale map of the Parish showing position of historical associations.

Register or Memorial of Ewell, Surrey edited by Cecil Deedes. This contains a chapter on the Topography of Ewell by Miss Glyn.

There is also a plan of the Parish of Ewell as in 1408.

Survey map of the Parish of Ewell of 1869, Scale 1/2,500. A photostat copy of this is available at Epsom Library. Sheets XIII/13A and XIII/13B are especially useful for the study of the Hogsmill Valley. Detailed plans of the mills are on these sheets.

17
Fieldwork and the Primary School

The purpose of this chapter is to describe some of the fieldwork studies undertaken by the pupils of a small country primary school. The children engaged in these studies were of mixed ability within the 7+ to 11+ age range. Whenever possible (it was not practical in all areas of investigation) the 6-year-olds were encouraged to become a part of the team and to contribute, in whatever way their age and ability allowed, to the study.

Not all the activities that made up these surveys are detailed below. For example the Village Survey included: (a) a flower collection, (b) an investigation into 'What we can buy in the village', (c) simple soil tests, (d) the parish council, (e) weather reports, (f) traffic counts, (g) water life in the brook.

Items (a), (b), and (g), were most actively supported by the younger children. The aim of the surveys was: i. to stimulate in the pupils an awareness and appreciation of the immediate surroundings and to arouse curiosity in that which lies beyond. ii. For the studies to act as a springboard from which to plunge into: surveys further afield, studies at greater depth of areas selected from the survey.

A VILLAGE SURVEY

1 Reason for the Survey

To discover all that we could about the people, their work, their recreation, the histories of the buildings, the changes that have taken place throughout time and, in particular, within living memory, the social facilities and everything that makes a village.

2 Preparation

(a) Obtain a 25" O.S. map of the area.

(b) If space allows, have this on display throughout (and after) the survey.

(c) Encourage map browsing and discussion on field names, where people live, where roads lead to, map symbols etc.

(d) Pupils make copies of the map.

(e) Practise elementary surveying techniques.

(f) Ensure that the required reference books are either in stock or available from the library service.

(g) If the survey is to be undertaken by pupils of a lower age group it may be necessary for the teacher to tape record extracts from some of the more adult reference books.

(h) Discuss with the class the scope of the survey, the proposed method of approach, presentation of results and individual and/or group activities.

(i) Seek the support of parents and others by asking for the loan of old photographs, newspaper cuttings, house plans and so on.

(j) Arrange for a party of pupils to visit the records office of the local newspaper.

(k) Investigate the possibility of inviting a local historian and others to the school.

3   Where People were Born

If the survey is to be undertaken within the confines of a small community it is quite probable that it will not be necessary to contact each individual as such information is often common knowledge.

(a) *Fieldwork:* prepare a master list on a street basis. Each pupil is made responsible for the details of his family. Mark off these as obtained. Allocate (on a street basis) the remaining population to groups. Information is obtained by house to house calls.

(b) *Presentation of results:* a very effective display may be built up by using small squares of coloured gummed paper. A square for each person. Use yellow to represent born in the county, blue for elsewhere in the British Isles and green for born overseas.

Using a large scale map (25″) of the area the squares are positioned where the people are now living. A map of the British Isles is mounted to the left of the area map and a World map on the right hand side. Run threads, secured by pins, from the coloured squares to place of birth.

The completed display may bring forth observations such as 'A lot of people were born near Hull. Why did they come here?' 'Where do the people who leave the village go?'

4   Where People Work

The distinction between 'where' geographically and 'where' in terms of the name of the employer is an important one. Very few people object to the geographical query but there is sometimes a reluctance to disclose information on the more personal question. The survey was concerned with the geographical 'where'.

(a) *Fieldwork:* information obtained by interviewing householders. If the person is not available mark N.A. against the house number and report to the pupil responsible for compiling the Recall List.

(b) *Presentation of results:* express your findings as a block graph. Discuss the reasons for the employment pattern. If a survey had been made 50 years ago would the pattern have been similar?

5   Cars and Vans owned by the People

(a) *Fieldwork:* it is often better to combine this activity with *Where the people were born* and/or *Where people work*, thereby saving the pupils' time and reducing possible inconvenience to the public to the minimum.

(b) *Presentation of results:* using a colour key pupils make a block graph of their findings.

6   How the Land is Used

(a) *Fieldwork:* after the initial survey made by groups who record crops and animals on a copy of the O.S. map changes are noted by individuals.

(b) *Presentation of results:* compile record of when and in which field crops are grown. Note sowing and harvesting dates. Record number and breeds of stock on pasture lands.

7   Bird and Animal Observations

(a) *Fieldwork:* not an organised activity. Pupils record observations made during other activities, to and from school and at weekends.

(b) *Presentation of results:* set up a tape recorder in the classroom. Pupils record observations. The report to follow a pre-arranged sequence: Name of observer; Date; Number and name of animal/bird sighted; Place. From reports compile permanent record.

8   Trees in the Village

(a) *Fieldwork:* the equipment needed – pencil, notebook, chalk mallet, plasticine. Although a number of pupils may be engaged in this study it is desirable that only one group of three or four, is in the field at any given period. If for example an oak is located it is i. given a number, ii. location recorded in note book, iii. a leaf and twig collected, iv. a bark impression taken.

When the groups return to school they report that an oak has been located and specimens collected. The out-going party is made aware of this and consequently duplication of effort is avoided.

To make a bark impression, i. Place slab of plasticine against the bark. ii. Pound firmly with mallet. iii. Chalk around plasticine. iv. Carefully peel off plasticine.

When the plaster cast has set it will be necessary to return to the tree to check on the colour – hence the chalk marking.

(b) *Presentation of Results:* i. record tree number on appropriate position on map. ii. Make plaster casts of leaf twig and bark. iii. Display on cards labelled with tree number and name. iv. Select tree for 'Study of an Individual Tree'. v. Measure the height, girth and foliage spread. vi. Observe and record dates of flowers forming, leaf bud bursting, tree bare etc. vii. Animals and plants that live in/on/around it.

9   Interviews

The objective of an interview can be either to obtain specific information or to give people the opportunity of talking on the subjects in which they are most interested. If specific information is required it may be necessary for a pre-interview discussion to be held.

People to interview: rural craftsmen, e.g. thatcher, blacksmith. Owners of old houses, landlords of inns, the vicar and older inhabitants of the village.

(a) *Fieldwork:* pupils contact people they wish to record on tape to fix time and date for interview. It is advisable for the pupils to mention that the discussions are to be tape recorded. The interviews are conducted by one or two pupils.

(b) *Presentation of results* tapes are filed for future reference.

10   The Village Offers You

It is often stated that nothing goes on in a village and this section of a survey tests the truth of the statement by discovering the range of sporting, social and welfare facilities available to the people.

(a) *Fieldwork:* collect information from notice boards, club fixture cards, time-tables and secretaries of various organisations.

(b) *Presentation of results:* print or duplicate a book containing full details of the facilities offered. Make this available to the public.

A Shopping Survey

1   Reason for the Survey

To investigate the shopping habits and needs of the public and to discover if the facilities offered by the nearest town were adequate.

2   Preparation

The success in a venture such as this, or indeed any other survey which depends on direct contact with the public, lies in the relationship established between the man in the street and the children. It is a case when the first impression is all important. It is therefore most advisable to give some thoughts to what is necessary to literally 'stop the public in their tracks' and to other points of organisation.

(a) The day of the week on which the survey is to take place. Select a day (such as market day) when there will be sufficient shoppers to ensure a representative sample and that the required numbers of interviews can be completed within the time schedule.

(b) The availability of public transport.

(c) Determine how many and where the interview points will be.

(d) Will it be necessary for more than one member of staff to accompany the party?

(e) Arrangements for mid-day meal.

(f) Discuss with children wording of form.

124

(g) Print or duplicate forms.

(h) The interview technique.

3   Interview Form

Frequently those stopped for interview read the form as it is completed. It is therefore essential that the form should be of the highest quality, be easy to follow and take only a minute or two to complete. A yes or no, a tick or a cross answer is to be preferred. Reduce writing by the pupil to a minimum.

A sample form.

Shopping Survey

Do you live in the town (A) or travel in (B)?
If (B) how far do you travel? How often do you come?
Please tick reasons for shopping here: cheaper
wider choice   social (meeting friends)   banking facilities
any other reasons
Are you satisfied with the service?   Is there sufficient
competition?   Is there a lack of any type of shop?

Do you use the supermarkets?   If yes please tick reasons:
price   easier   choice of product   trading stamps
other reasons
Do you collect trading stamps?   Would you prefer price
reductions to stamps?   Is there anything (e.g. furniture)
that you go elsewhere to buy?
Do you have a weekly delivery of groceries?
Is there a need for more delivery services?

The information from the completed forms is then transferred to binary punch cards. A card is made for each person interviewed.

4   Binary Punch Cards

The key card should be planned out at the same time as the questionnaire is compiled. Each hole represents the answer to a question. A slotted out hole indicates a YES reply. If the answer is NO the hole is left complete.

A hole is not needed for (A) (1st question on form) as a complete B hole indicates that the person DOES NOT come in and therefore must be (A).

To reduce the number of holes required the 'How often do you come?' holes are combined to once and twice a week on the same hole and similarly three and four times a week on the next hole.

(a) *Information recorded on card:* this shopper comes in, travels 3½ miles, shops once a week. She shops in the town because it is cheaper, for social reasons and because it is

FIG. 41

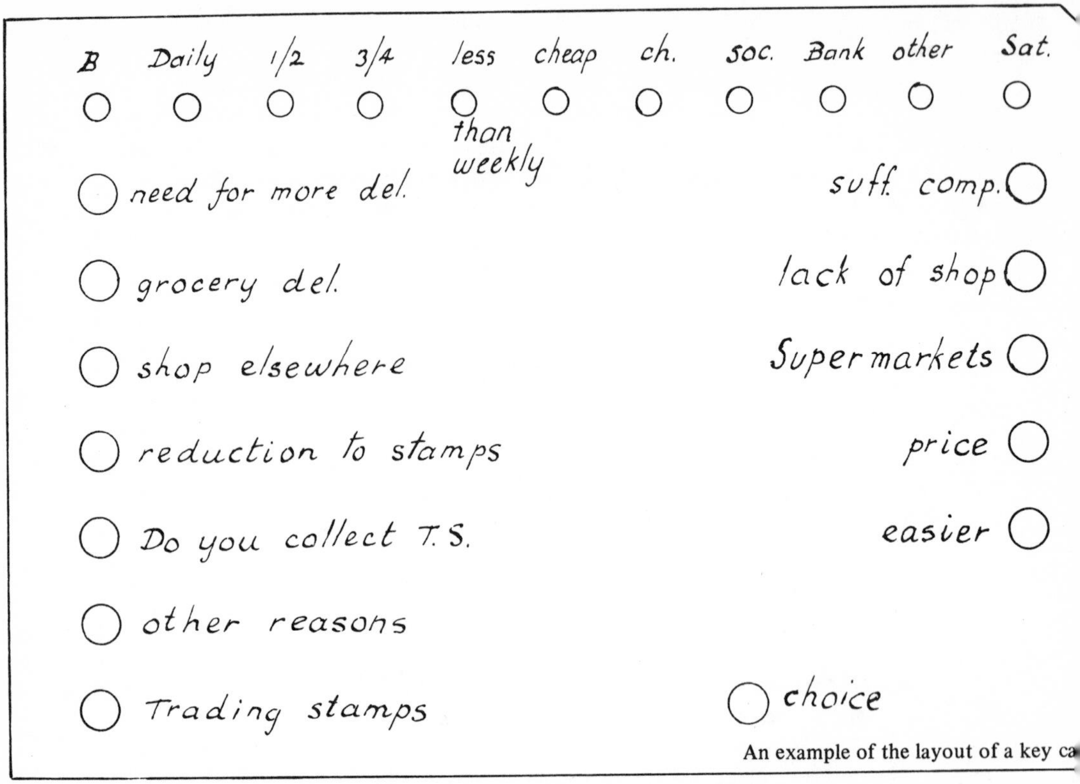

B	Daily	1/2	3/4	less	cheap	ch.	soc.	Bank	other	Sat.
○	○	○	○	○	○	○	○	○	○	○

than weekly

○ need for more del.   suff. comp. ○

○ grocery del.   lack of shop ○

○ shop elsewhere   Supermarkets ○

○ reduction to stamps   price ○

○ Do you collect T.S.   easier ○

○ other reasons

○ Trading stamps   ○ choice

An example of the layout of a key ca

convenient. She is satisfied with the service, believes that (in general) there is sufficient competition but there is a lack of departmental stores. She uses the supermarkets because of the choice of products available. Would prefer price reduction to trading stamps. Goes elsewhere to buy clothes.

N.B. Additional information such as type of shop needed is written on the card next to the appropriate hole.

(b) *Method of sorting:* we require to know how many of the shoppers who come into town do so to use the supermarkets and if they believe the supermarkets offer a wider choice of products.

i. Stack cards. The cut off corners show that the cards are correctly positioned.
ii. Insert a thin steel needle through the B hole. Lift the stack of cards. Cards with B hole slotted out will fall out. These are all the shoppers who came in.
iii. Collect up these B cards and re-stack.
iv. Insert the needle through the supermarkets hole. Lift, and slotted cards will fall out. These are the shoppers who come in and use the supermarkets.
v. Collect cards and re-stack.
vi. Insert the needle through choice hole. Lift, some cards fall out. Sorting completed.

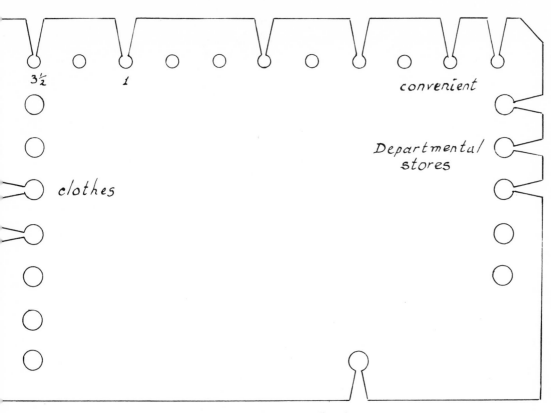

FIG. 42 A completed "key" card

(c) *The analysis:* using the punch cards, prepare an analysis and print or duplicate copies for distribution.

(d) *An example analysis:* 100 shoppers were interviewed – 48 lived in the town and 52 came in. 46 had to travel under 10 miles, 5 between 10–20 miles and 1 a 100 miles. The frequency of visits varied from daily to once in 16 months. 19 of the 48 locals shopped daily and 15 twice a week. 67 per cent of those who travelled in shopped once or twice a week. The reasons for shopping here were: cheaper, 31. Wider choice, 31. Social, 28. Banking facilities, 18. Other reasons (e.g. work here, nowhere else to go, 35.91 per cent were satisfied with the service given and 85 per cent thought that there was sufficient competition. 63 per cent stated there was a lack of some type of shop, 44 suggested a need for a drapery, dress stores etc. 12 named a departmental store.

90 per cent used the supermarkets for the following reasons: Price, 31. Easier, 33. Wider choice, 35. Trading stamps, 0. Other reasons, 3. Only 19 per cent collected trading stamps. 92 per cent would prefer price reduction to stamps, 7 voted for stamps and 1 was indifferent. 62 per cent shopped elsewhere for some things (18 for clothes,

10 for furniture). 20 of the 48 locals and 25 of the 52 who came in have a weekly delivery of groceries. 21 locals and 12 out of town shoppers believed there was a need for more delivery services.

The Forest Plot

The Adoption of a Forest Plot Scheme introduced by the Forestry Commission enables individual schools to adopt a forest plot. The procedure is a simple one. Schools wishing to take part in the scheme should write to the local Conservator of Forests who, provided that he is satisfied that there will be a sustained interest, will arrange a meeting with the District Officer to select a site. Following this the only requirement is that the school or L.E.A. sign a form of indemnity.

Before an approach is made to the Conservator consideration needs to be given to some factors on which the success of the scheme depends.

(a) Is the school within a reasonable distance from one of the Commission's forests?

(b) If it is how is the cost of the journey there to be met?

(c) Do the proposed transport times fit in with school buses?

(d) How frequent will the visits be?

(e) The number and age range of the pupils to be taken.

(f) The staff/pupil ratio required.

(g) Can these requirements be met within the existing school organisation? If not what changes are desirable and are these possible?

Assuming that the conditions governing the decision to adopt can be met and the go-ahead is given the probable work plan will be:

1 Preparation of the land

This varies with the general condition of the plot, but will most likely be the clearing of bramble and ferns, the removal of coppice shoots from old tree stumps and the stacking of unwanted logs and branches.

2 Planting

Before the start of the planting season it is necessary to draw up a planting cycle plan. A four or five year period is suitable for an acre plot. This allows the pupils to complete the actual manual forestry work and leave ample time for other pursuits.
The usual spacing at which the trees are set are 6ft $\times$ 6ft, 5ft $\times$ 5ft or 4½ft. A 6ft $\times$ 6ft planting gives 1,210 trees per acre. The planting lines are sighted using three poles.

3 Weeding

Throughout the summer terms of the first few years after planting frequent weeding will be necessary to prevent the smothering of the young trees. This applies to all sites, but particularly so if the trees are planted amongst bracken and brambles.

4   Beating-up

However carefully the planting is done some losses will occur. New trees must be planted in the gaps. This is 'beating-up' and takes place a year after planting.

Tools used for these operations are: garden spade, reap hook, bill hook, bow saw and felling axe. In addition to the basic forestry work a plot offers the opportunity for experience in a wide range of activities.

(a) *Mapping:* this need not necessarily be restricted to the plot (although this will provide a useful exercise) but should extend to the whole forestry area. The school plot is sited in 165 acres of woodland.

(b) *Bird life:* a start with the identification of species may lead to a bird census, the making of nest boxes and the recording of bird songs.

(c) *Animal tracks:* a collection of casts of footprints can be built up. The date and position of tracks can be recorded on a map.

5   Bringing the Forest into the Classroom

There is no need for the activities stimulated by a woodland to be limited to the time spent on the site. Back in the classroom a twisting root, a charred log salvaged from a clearing-up fire, a random branch, or in fact any piece of wood that takes on a significant shape can be worked on (using surform tools and sandpaper) to bring out the shape. At a later date pupils may wish to use gouges to emphasise line.

The use of coppice tools and devices enables the children to experiment in the practical (in addition to the purely artistic) usage of wood. When a hazel, sweet chestnut or ash is felled the stump (stool or moot) will remain alive in the ground for many years. This will send out new growth. It is from this that coppice material is obtained. The making of stools and small chairs is well within the capabilities of junior children.

The tools necessary to introduce this craft are a frame saw, a froe and mallet (for cleaving) a side axe, a draw knife and a set of rounders. Suggested reading: *Tools and Devices for Coppice Crafts,* Young Farmers Booklet, No. 31.

Conclusion

A consensus of opinion as to the problems involved in the organisation of fieldwork suggests that the difficulties fit into two categories, financial and staff/pupil ratio. Unfortunately neither of these can be solved by set formulae.

(a) Financial: this may be more pronounced in the Primary School where recognition by the public of the place of fieldwork in the school programme is not always so readily given.

Transport costs may be met from i. capitation, ii. school funds, iii. parent contribution.

(b) *Staff/pupil ratio* required is governed by i. the nature and place of the activity, ii. the age and experience of the pupils taking part, iii. the duration of the activity.

At infant and lower junior level a ratio of absolute maximum: 1−20, desirable: 1−15, overnight trips: 1−10, is advisable.

Very few, if any, schools, staffing can allow this ratio without a considerable alteration of programme and a willingness of colleagues to take on added responsibilities. It may well be that the solution to this lies in a greater co-operation between the schools and the colleges of education many of which have expressed their willingness to release students from college activities so that they could assist schools in fieldwork programmes.

Further Survey

One of the stated aims of these surveys was for the studies to act as a springboard from which to plunge into 1. surveys further afield, 2. studies at greater depth of areas selected from a survey. At Primary level 1. is often more appropriate.

Participation in the surveys has stimulated suggestions for further investigations.

(a) *A Footpath Survey:* to map and report on the condition of the footpaths in the area.

(b) *A Public Transport Survey:* to discover if the services are sufficient for shopping, social and working needs.

(c) *A River Survey:* a study of selected stretches from the source to the mouth.

(d) *A Canal Study:* to discover the use made of the inland waterway system.

Much of that which contributes to the success of fieldwork is common to success in all aspects of school life and the attitudes developed in the field are not confined and should be spread to influence the children's approach to learning in all subject areas. Experience in fieldwork tends to develop. i. self-discipline, ii. confidence to approach the public and new learning situations with self assurance, iii. power of observation, iv. the ability to ask yourself questions and make decisions, v. realisation that learning situations exist beyond the school building.

18
Fieldwork and the Young School Leaver

The young school leaver is characterised as one who is impatient with the confines placed on him by the somewhat academic schooling he is often obliged to follow. He looks forward eagerly to taking his place in the world of work, and making his own way in life. The Schools Council Working Paper No. 11: *Society and the Young School Leaver*, makes the important point that, 'acute needs in terms of personal development, can never be satisfied by a curriculum based on a textbook. What the pupil needs is the chance to learn more about himself, and about the community in which he lives'. Clearly, the use of fieldwork is an eminent teaching method for helping to satisfy many aspects of these pressing needs. This was recognised some years ago by the late Professor Wooldridge in his 1954 Presidential Address to the Geographical Association entitled: 'The State of Geography and the Role of Fieldwork', which he concluded with this parody on a Wordsworth poem:

> 'One traverse in a Surrey Vale
> (Or if you prefer it Yorkshire Dale)
> Will teach you more of Man,
> Of man in his terrestrial home,
> Than all the textbooks can!'

Of course, to this one might add that a traverse in an urban area can provide the same kind of educational experience. Nevertheless, there is a need for caution in proclaiming the virtues of geographical fieldwork because fieldwork itself can become a sterile activity. The geography teacher can find himself applying techniques rather than placing his children in learning situations. For instance, the value of carrying out a traverse is limited, indeed, if it forces the young fieldworker to look for correlations in rock, soil, vegetation etc. which may only exist in the neat abstractions drawn in textbooks. Equally, geographical fieldwork with children can come to defeat its purpose if the teacher has to draw the line between what is geography and what is not geography. The apocryphal story is told of one geography teacher leading a party on a village study. On coming to the village church he turned to his pupils and said, 'We won't go inside there as it does not contain any geography!' But the fact is a purist approach to what constitutes geographical fieldwork is largely irrelevant to the educational needs of most children, and in particular to those of the young school leaver.

In this connection it is interesting to note that no examination of the curriculum objectives of fieldwork has been attempted, as far as is known to the author. One pioneering experiment carried out by the Geography Department of the City of Leicester College of Education, indicated that there might be little difference in the amount of information acquired as a result of fieldwork compared with that obtained from class teaching dealing with the area visited. Rather, it would appear that the value of fieldwork is the effect it has on the patterns of learning behaviour, such as the awakening of curiosity, the increased ability to follow-up 'clues' in the solution of problems encountered during the fieldwork investigations, the stimulating contact with people and things during the excursion, and so on. 'Going out and discovering for oneself' is surely the essence of fieldwork, and the trail of discovery cannot be made to stop at a subject boundary. In particular, the teacher must be careful that he does not impose on the pupil the same kind of limitation implicit in the use of a textbook, where the knowing of facts is more important than the process by which we find out about facts.

It is important too, that the work in the classroom should be closely united with the work in the field. The follow-up to the fieldwork and the preparation for it should be in a practical form, so that classwork becomes an active enquiry through the use of documents, reference material, visual aids, specimens and discussions, and the geography room is considered as a laboratory where practical investigations can be carried out. With this aim in view case studies can be selected which embrace both fieldwork and classwork, and which might well stray beyond the confines of geography as an academic discipline. The Working Paper No. 11 provides useful criteria for selecting such areas of enquiry, which are as follows:

(a) 'It is essential that the start of the topic should be interesting to the pupil now at this point in their lives.

(b) It should also be, in the opinion of the teacher, relevant to their future and of lasting significance.

(c) The topic should be capable of stimulating the pupils to personal involvement, creative thinking and individual curiosity.

(d) It is desirable that the topic be exploitable in the local situation. The resources of the school and the neighbourhood may often suggest the selection of one rather than the other of two topics of apparently equal worth.

(e) It is more likely to be successful if the topic fits in with the particular specialisms, enthusiasms and qualifications of the teachers concerned, and if it can be backed by adequate material made locally or obtained from a central project.

(f) It is sensible to keep a reasonable balance between the various branches of human knowledge.'

The report prepared by the Geographical Association, 'Geography and the Raising of the School Leaving Age' (1966), also suggests cogent ways and means for tackling this task. It emphasises how the course for these pupils needs planning in relation to their physical, technical, and social environment, that their work should develop individual practical activity leading to out-of-school investigations, and that team teaching can take place, hand-in-hand with teachers of other subjects, in working out a series of combined studies. The report recognises that not all aspects of such a course

132

is geography, 'but the geographer contributes many essential elements, and is probably the best equipped to provide "bridges" between traditional subject departments'. That this is so is evidenced by the fact that of the eleven areas of enquiry suggested by Working Paper No. 11, eight of them strongly involve the geography teacher.

Both the Schools Council and the Geographical Association Report emphasise the need to use the local environment as the source of learning situations for these pupils. The former states that 'much of the successful work has come from schools which decided to use the potential of their local environment'. This is very true, nevertheless an inbuilt danger does exist that the pupils may become too parochial in their fields of enquiry. To counterbalance this possible trend, the geographer has a useful teaching approach at his command, the sample study method, which if applied as an outcome of the work in the local environment can be of great value and interest and lead out to the world beyond the immediate locality. As Professor R. C. Honeybone once pointed out, 'The sample study by using field study in the classroom enables the pupil to gather information for himself about the particular sample'. For instance, an investigation into the life of immigrants in the neighbourhood might well lead on to a sample study of Calcutta. This would be a sample study which would extend as much into the historical and sociological aspects of the urban problem of Calcutta as it would deal with its geography.

It is the purpose of the following two examples to indicate the kind of contribution the geography teacher, basically using the fieldwork and practical approach outlined above, can make to the creation of a suitable course for the young school leaver.

1 Man and the Motor Car

The Geographical Association Report argues that 'pupils . . . ought to know something of planning in modern society and geography teaching contributes directly to their understanding of this'. This is a particularly valid point when it is considered that at least eighty per cent of our children live and are taught in an urban environment. Urbanisation is increasingly affecting life and landscape in this country – as in the world generally. But the teacher's problem is not to impose this interest on the pupils, but to build on an interest they already have in their urban surroundings.

In this way, 'Man and the Motor Car' provides an area of enquiry which can be linked with work in the class and in the field, and which cuts across the traditional subject boundaries. It is also a theme which connects directly with the interests and daily lives of these pupils if only because many of the boys will be hoping shortly to ride their own motorbikes and are often fascinated by the workings of the internal combustion engine. Even the girls do not lack an interest in the subject – many of them are potential pillion passengers.

The geography teacher's starting point might be to consider the geography of car parking as a way of stimulating an interest in the wider aspects of the topic. No attempt is made here to provide a formula for lesson preparation, but a series of suggestions are offered which attempt to satisfy the requirements noted above. The investigations involve the practical use of maps of various kinds met within daily life, the use of reference material generally available in the public libraries, face-to-face enquiries with officials and members of the public, and the use of photographs, diagrams and tape recordings in presenting the results of the enquiry. This work has been developed for use in Leicester, but it might be attempted equally in other urban areas.

(a) *Introduction:* a useful lead lesson for a team teaching approach can be a discussion based on the show of the film *'Traffic in Towns'*. This is the film of the Buchanan Traffic Report and is produced by the Central Office of Information. Associated class reading can be taken from *'Urban Growth in Britain'* by Michael Storm (Oxford University Press). This book contains many suggestions for practical and fieldwork investigations in urban areas.

(b) *Directions to Pupils:*
i. Obtain a copy of the car parking pamphlet available free from the Information Bureau, and also consult out-of-date pamphlets that may be retained in the Central Reference Library. From these make a map to show the present position of car parks in Central Leicester, and on a tracing overlay show the former situation of car parks now closed or built over. Enquire of Local Government officials the reasons for any changes that might hhave occurred.
ii. Interview members of the school staff, and as many parents as possible, to discover the parking places they most frequently use when visiting the centre of the town. Plot these on a second overlay to be placed over the maps in i. above. What differences do you notice in the parking patterns shown on the three maps? Try to discover the various reasons for using, or not using, the official car parks.
iii. From information obtained from the car park attendants, draw up column graphs to show the car parking capacity of the present car parks.
iv. On a map of central Leicester mark in black all those areas serving the motor car exclusively, e.g. streets, garages, sales, parking spaces, petrol stations, etc. and then estimate approximately the total area covered by these.
v. By interviewing traffic wardens find out what regulations have been made to control car parking in the centre of the city.
vi. Locate on the base map that you drew for (a) above the multi-storey car parks situated in Leicester. Is there a pattern discernible in their distribution? What reasons might there be for their siting? (e.g. their nearness to the main shopping centre, or distance from the mainline railway station.) What charges for parking do they make?
vii. To what extent does the design and method of construction of the multi-storey car park improve or spoil the townscape? Take photographs to illustrate your opinion.
viii. Study the view over the city from the top of the multi-storey car park. Use the view finder method described on page 18. Name the major landmarks in view, such as church spires, point blocks, factory chimneys, departmental stores, etc. Compare the view with air photographs of the immediate area. Have any changes taken place since the photograph was taken? How can these be explained?
ix. Find out what buildings were cleared away to make room for the building of the multi-storey car parks. This can be done by referring to the older Ordnance Survey maps (25" to the mile) kept in the Central Reference Library, and by consulting a pre-war copy of *Kelly's Street Directory*.
x. Make a survey of the shops built under one of the multi-storey car parks. What

types of shop are found here? How do they differ from shops in a normal shopping area? What conclusions can be drawn from this survey?

xi. Ask shopkeepers, and a number of pedestrians in the area, their opinion of the effectiveness of the multi-storey car park. To what extent is it solving the car parking problem in the area? It is attracting more or less trade to the area?

xii. Interview drivers of vehicles using the multi-storey car park in order to ascertain 1. their reason for using the multi-storey car park and not other places for parking; 2. the place of origin of their journey; 3. their opinion of the car parking arrangements in the centre of the city. A portable tape recorder could be used for these interviews.

xiii. Using the key for car registration numbers given in the *A.A. Handbook*, find out the place of origin of the vehicles as shown by their number plates. Draw a map to illustrate the results of this investigation, and compare with a map drawn to illustrate ii. above. Discuss the limitations of this method.

xiv. Find out from the Report *Traffic in Leicester* by Conrad Smigielski the plans for car parking in the Leicester of the future, e.g. the reasons for interchange car parks, plans for overcoming congestion etc.

(c) *Follow-up:*

i. a display for the work achieved.

ii. Project work on the motor car industry involving a visit to a car factory; the history of the motor car, leading to model making and museum visits; lessons based on the practical maintenance of petrol engines; discussions on the impact of the car on society; a sample study of Los Angeles.

(d) *Further References:*

i. *Parking Matters: a Community Problem*, a Roads Campaign Council Publication. A good book on which to centre discussion, giving brief accounts of the way the major provincial cities are tackling the problems of car parking.

ii. *Cars and Car Owners in Leigh*, University of Manchester, Department of Extra Mural Studies. A useful guide to the making of the sociological type of survey.

iii. *Planning for Man and Motor* by Paul Ritter, Pergamon Press. 1964. A useful advanced text for detailed reference.

iv. A list of films covering many aspects of the automobile and its use for business and pleasure were given in the Times Educational Supplement for 8 April 1966.

2 Patterns of Industry

In the Newsom Report, and in the publications of the School Council, there is much importance given to the necessity of introducing a vocational interest into the education of the young school leaver. Clearly, an important aspect of this kind of approach is to carry out visits to factories and other places of work in the school area. Here again the geographer can make a strong contribution, particularly because much geography can be derived from the study of a film. This kind of investigation can reinforce the experience of the work environment given by the visit, and encourage a better understanding of the type of industries and their problems which the firm may represent.

The working Paper No. 11 gives a suggestion for a related area of classroom enquiry which it entitles 'The Triangle of Prosperity'. The aim of this enquiry is to 'show the present advantages of industrial development within the triangle of Birmingham-Cologne-Paris'. A number of useful suggestions are made for carrying out this work, such as 'small groups can take individual industries within the triangle and draw maps and diagrams showing where the raw materials come from, and where the finished product is sent'. But such work might well be an abstract exercise unless the pupil has visited and experienced a factory environment at first hand.

Furthermore, there is a need to examine the prospects of employment for a young school leaver, taking his own locality as the potential area for work. This may start from an examination of the current employment figures published in the newspapers, and lead on to a consideration of the nationwide factors such as are illustrated in the article, 'Opportunity and Affluence' by E. M. Rawstron and B. E. Coates (*Geography*, January 1966). This might develop into studies of employment during the inter-war Depression and lead on to an investigation of Economic Development Council plans for the home area. (These plans, issued as reports, as is the case for many of the town planning publications, are frequently rich sources of information that can be used in geography teaching.) This work can be linked with a case study of a contrasting area in Britain. A way of doing this is for the teacher to collect 'banks' of articles and pictures appearing in the national press. As these are assembled they can be stored in cardboard boxes or envelopes of suitable size, and labelled accordingly. Material collected in this way has to be sifted and edited by the teacher, but as far as is possible the original articles should be made available to the pupils. A small group of pupils then investigate each area for which information has been collected, seeking answers to a prepared questionnaire. For instance, a group could be placed in the position of a manager of a firm choosing the best place to locate his firm. One of the articles would be read by one of the pupils in the group, and then the group of pupils would pool the information they have gathered to answer the questionnaire.

(a) *The Geography of a Factory*

It is an essential part of a factory study to examine the manufacturing process which is taking place there. This is often quite complex, and unless the pupils are well-briefed beforehand they will not necessarily understand what is going on when they make their visit. When possible, the teacher should prepare a plan of the factory, and a flow diagram to illustrate the sequel of operations involved in the factory process.

Preparatory fieldwork by the pupils involves mapping the land use of the area in which the factory is situated. The method of carrying out a factory estate survey is explained below. By way of contrast a survey of manufacturing in a mixed urban area might also be carried out along these lines:

i. Plot on a base map all the factories, warehouses, workshops, and suppliers to the trade within a suitably selected area of the city. (On this finished map this can easily be done by using coloured dots punched out from sticky-backed paper.) If any form of zonal grouping emerges, indicate this on the map.

ii. Investigate as many representative examples of the firms as possible to establish the reasons for their location. This involves preparing a suitable questionnaire which could be sent by post, or used in interviewing a sample of the firms.

136

iii. With reference to the *Kelly's Postal Directories* or Trade Census Returns available in the Central Reference Library, plot on a base map the location of firms in the same area on selected dates in the past. Discover changes that have taken place and try to account for these. Is there any evidence of firms moving in or out of the area within recent years?

iv. Consult the Town Development Plan. What is the future plan for the area under investigation?

v. Display the maps and information that is gathered in this way, illustrated by photographs, labelled examples of information obtained from the firms such as brochures, sample of products and raw materials used etc.

(b)   *The Factory Survey*

This may be undertaken in one of the following ways:

i. By the teacher alone who wishes to acquire information, experience, and illustration for presentation in the classroom. Particularly useful, for instance, is to take a series of colour slides, where possible, to use later in the classroom to illustrate the stages in the manufacturing process.

ii. By the pupils on a visit organised by the teacher so as to make a detailed survey by finding the answers to a prepared questionnaire. Obviously, this kind of work is best suited for a factory where the process involved is simple and the conditions are such that the pupils can make a close study.

iii. By a smaller group of pupils who can be given access to documentary material and other information (see below) from which they can plot maps of various kinds.

(c) *The Factory Survey Questionnaire*

It is not possible to plan a questionnaire which can be applicable to all types of factory. But the following example demonstrates the kind of questions that can usually be included.

Name of pupil _____ Date of Survey _____

Name and address of firm _____

When was the firm founded? _____

Was it founded here or on another site? _____

What previous firms have occupied the present site? _____

What were the special advantages of the site in the past? _____

Are these factors as important today as they were? _____

At about what period did conditions change? _____

What is the source of water? _____

What is the source of power? _____

What goods does the firm produce? _____

Are there any by-products? _____

If so, list them here _____

The following are the stages of production. Describe them as you go round the factory, collecting where possible samples to illustrate each stage in the process of production.

FIG. 43

FIRM STUDY OF INVICTA PLASTICS LTD. (1)

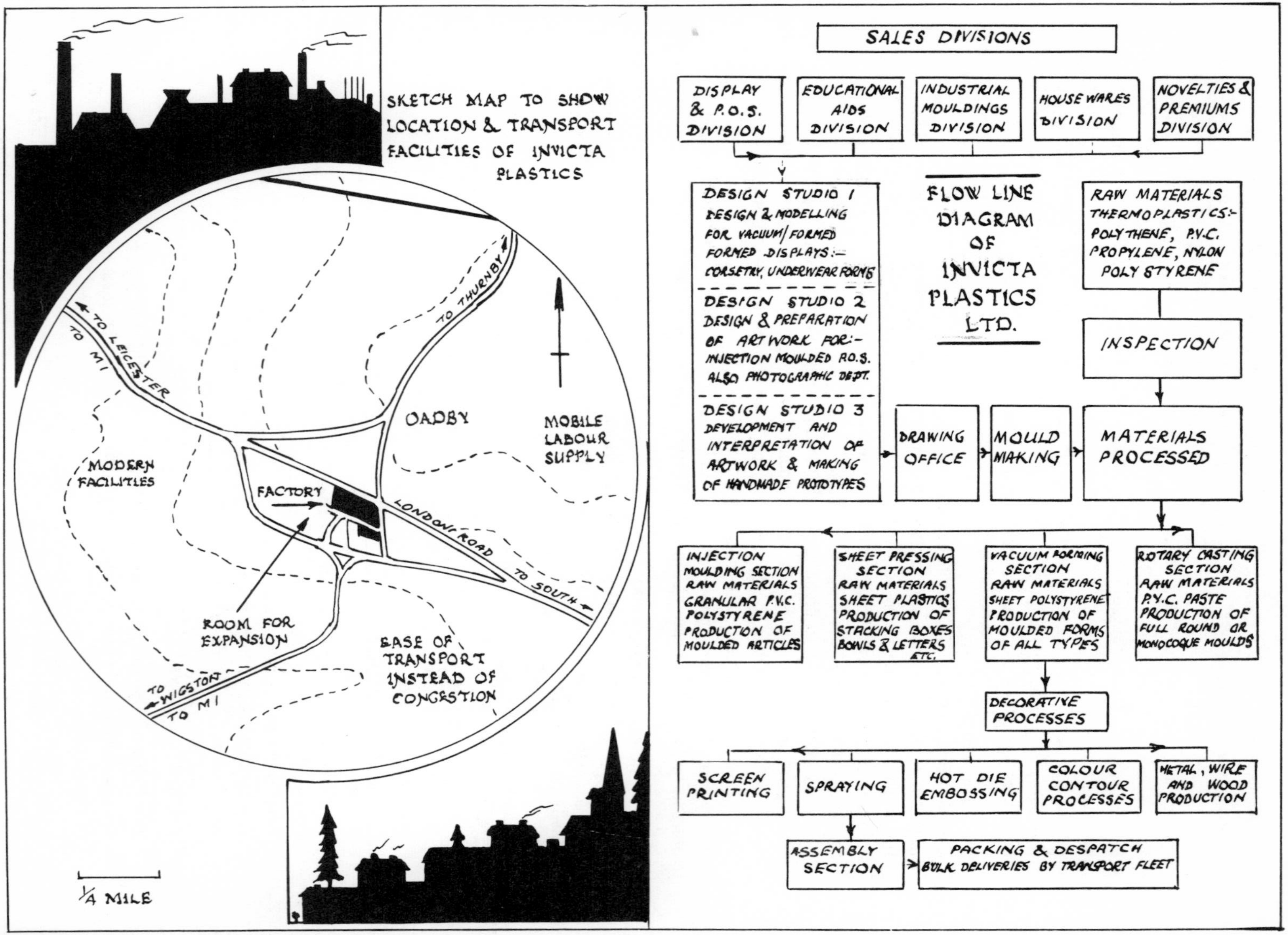

FLOW LINE DIAGRAM OF INVICTA PLASTICS LTD.

SALES DIVISIONS

- DISPLAY & P.O.S. DIVISION
- EDUCATIONAL AIDS DIVISION
- INDUSTRIAL MOULDINGS DIVISION
- HOUSE WARES DIVISION
- NOVELTIES & PREMIUMS DIVISION

DESIGN STUDIO 1
DESIGN & MODELLING FOR VACUUM/FORMED FORMED DISPLAYS:— COASETRY, UNDERWEAR FRONTS

DESIGN STUDIO 2
DESIGN & PREPARATION OF ARTWORK FOR INJECTION MOULDED P.O.S. ALSO PHOTOGRAPHIC DEPT.

DESIGN STUDIO 3
DEVELOPMENT AND INTERPRETATION OF ARTWORK & MAKING OF HANDMADE PROTOTYPES

DRAWING OFFICE

MOULD MAKING

RAW MATERIALS THERMOPLASTICS:— POLYTHENE, P.V.C. PROPYLENE, NYLON POLY STYRENE

INSPECTION

MATERIALS PROCESSED

INJECTION MOULDING SECTION
RAW MATERIALS GRANULAR P.V.C. POLYSTYRENE PRODUCTION OF MOULDED ARTICLES

SHEET PRESSING SECTION
RAW MATERIALS SHEET PLASTICS PRODUCTION OF STACKING BOXES BOWLS & LETTERS ETC.

VACUUM FORMING SECTION
RAW MATERIALS SHEET POLYSTYRENE PRODUCTION OF MOULDED FORMS OF ALL TYPES

ROTARY CASTING SECTION
RAW MATERIALS P.V.C. PASTE PRODUCTION OF FULL ROUND OR MONOCOQUE MOULDS

DECORATIVE PROCESSES

- SCREEN PRINTING
- SPRAYING
- HOT DIE EMBOSSING
- COLOUR CONTOUR PROCESSES
- METAL, WIRE AND WOOD PRODUCTION

ASSEMBLY SECTION

PACKING & DESPATCH
BULK DELIVERIES BY TRANSPORT FLEET

SKETCH MAP TO SHOW LOCATION & TRANSPORT FACILITIES OF INVICTA PLASTICS

TO THURNBY

OADBY

MOBILE LABOUR SUPPLY

LONDON ROAD

TO SOUTH

EASE OF TRANSPORT INSTEAD OF CONGESTION

FACTORY

ROOM FOR EXPANSION

TO LEICESTER
TO M1

TO LEICESTER
TO M1

MODERN FACILITIES

¼ MILE

First of all _____

Then _____

Next _____

After this _____

The fifth stage in the production is _____

After this has been completed _____

Later on _____

Finally _____

What is the average production of the factory:

Daily _____ weekly _____ annually _____

What kinds of things might interrupt production? _____

Make a list of the raw materials used in the factory and where each comes from:

_____

Make a list of the places where the products of the factory are sold _____

Note any special orders that the factory has undertaken _____

How many people work at the factory? _____

What is the management organisation of the factory? _____

What special facilities are there for the employees? _____

What effect does the factory have on the locality? _____

Waste materials _____ Transport arrangements _____

Smoke _____ Housing _____

Water pollution _____ Employment _____

(d) *Making a map study of the factory*

Dr E. M. Yates and M. F. Robertson comment in their article, 'Geographical Field Studies' (*Geography*, January 1968): 'The danger lies in the fact that the works visit may become as easily non-participating as the mostly badly led conventional guided tour'. This danger is a real one, as with all fieldwork, but it should not occur if the visit is organised as part of a planned, practical course, and the questionnaire is one that can be really used during the visit. In addition, the following enquiries, requiring presentation by maps and diagrams, can also be made, (see Figures 43 and 44), and the results of this work clearly demonstrate the geography to be discovered in a study of a firm i. A plan of the factory lay-out. ii. A map to show the location of the factory, and its transport connections. iii. A map to show the source of the raw materials. iv. A map to show the destination of the products in the home market and abroad. v. A map to show the location of other firms linked with the one being studied. vi. A graph illustrating the production figures over a suitable period of time. vii. A distribution map to show the home locations of the employees. viii. A pie diagram to show the proportion of male and female employees.

In order to carry out this investigation a small group of pupils should be sent to the factory, and several factories typical of the school locality can be studied in this way. The resulting maps and diagrams can be made into an Atlas of Local Industry.

3 **Fieldwork and Interviewing**

Another useful form of fieldwork is to send children out to interview people in order

FIG. 44   FIRM STUDY OF INVICTA PLASTICS LTD. (2)

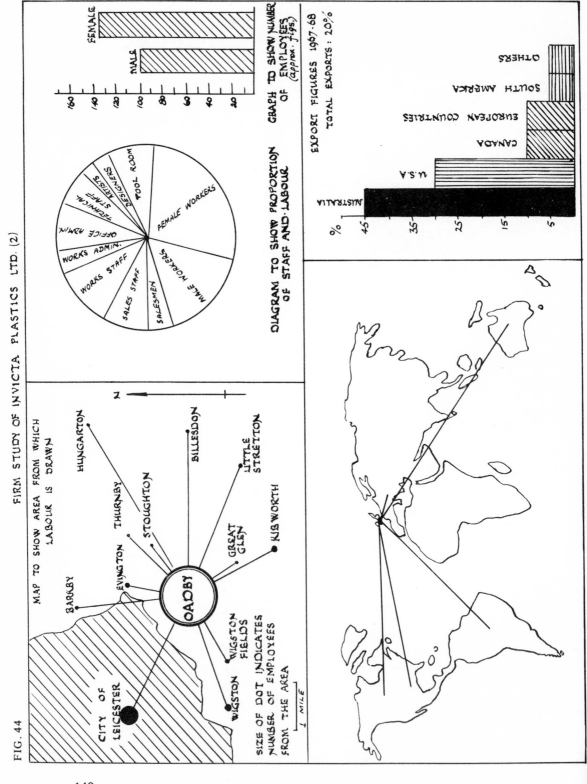

MAP TO SHOW AREA FROM WHICH
LABOUR IS DRAWN

N

CITY OF LEICESTER

BARKBY
EVINGTON
THURNBY
HUNGARTON
STOUGHTON
BILLESDON
LITTLE STRETTON
GREAT GLEN
KIBWORTH
OADBY
WIGSTON FIELDS
WIGSTON

SIZE OF DOT INDICATES
NUMBER OF EMPLOYEES
FROM THE AREA

1 MILE

GRAPH TO SHOW NUMBER
OF EMPLOYEES
(approx. figs.)

FEMALE
MALE

160 140 120 100 80 60 40 20

DIAGRAM TO SHOW PROPORTION
OF STAFF AND LABOUR

FEMALE WORKERS
MALE WORKERS
SALESMEN
SALES STAFF
WORKS STAFF
WORKS ADMIN.
OFFICE ADMIN.
TECHNICAL STAFF
DESIGNERS
ARTISTS
TOOL ROOM

EXPORT FIGURES 1967-68
TOTAL EXPORTS: 20%

AUSTRALIA
U.S.A
CANADA
EUROPEAN COUNTRIES
SOUTH AMERICA
OTHERS

% 45 35 25 15 5

140

to gather opinions and information concerning a project. This kind of enquiry needs careful organisation as well as tact, and the people to be interviewed should be approached by the teacher beforehand in order to secure their co-operation.

The questionnaire should be drawn up in collaboration with the pupils, so that the purpose and meaning of the enquiry is fully understood. When this task has been completed, the questionnaire should be neatly produced and phrased in good English. The class should practice the technique of approaching people for information. This is as much a case of English in action as it is Geography! And it can be linked with free dramatisation and role-playing exercises.

J. N. Jackson in *Surveys for Town and Country Planning* (Hutchinson University Press) gives instructions in the method for making surveys of this kind. Such enquiries can also be linked with the official methods of gaining information from the public to form an 'area of enquiry'. In fact, schoolchildren could contribute in this way to quite useful investigations applicable to planning or sociological enquiries. This is well demonstrated by P. N. Grimshaw in his article on 'Population Mobility – a Case Study' (*Journal of the Town Planning Institute*, March 1968) which describes an interesting investigation into an aspect of Social Geography carried out with the help of schoolchildren.

In certain circumstances, the use of the portable tape recorder is helpful for this kind of work. Interviewing older members of the community in order to gain information of the appearance of the locality in the past is the kind of fieldwork that could be attempted.

19

The Study of a River

The study of a river is one of the most popular kinds of fieldwork enquiries with schoolchildren. This is partly because the student comes near to seeing the modelling of the landscape in action, as it were; also simply because it is fun to be alongside water.

The amount of theoretical preparation required will depend upon the age and level of attainment of the children concerned. But something about river action can be usefully taught to children in the top classes of the primary school as well as to the sixth form at the grammar school. The depth at which this study is taken varies, of course, from the simple idea of 'wearing away' the land to the higher flights of geomorphological interpretation required for A level geography. But if it is well taught by means of lively practical exposition in the field, even fourth-year leavers in a secondary modern school can return breathless from a fishing expedition to tell their teacher how they found themselves sitting near a river terrace!

Any small stream or gulley makes a good and simple starting-point. And these can be found in town parks; in accessible sandpits, as well as in country districts. Alternatively the teacher can prepare a demonstration for his class using a large heap of earth and a hosepipe. By this means he can show that a stream acquires a load of sediment in its upper reaches by deepening its channel and then transports this load downstream. Lower down the artificial stream forms meanders where undercutting may take place on the outside of the bends to make miniature 'rivercliffs' (see Figure 45). A meander pattern will become more pronounced as the banks retreat at the undercut rivercliffs, and in the lowest reaches of the stream deposition becomes all-important; braiding and delta formations can be made to occur.

FIG. 45

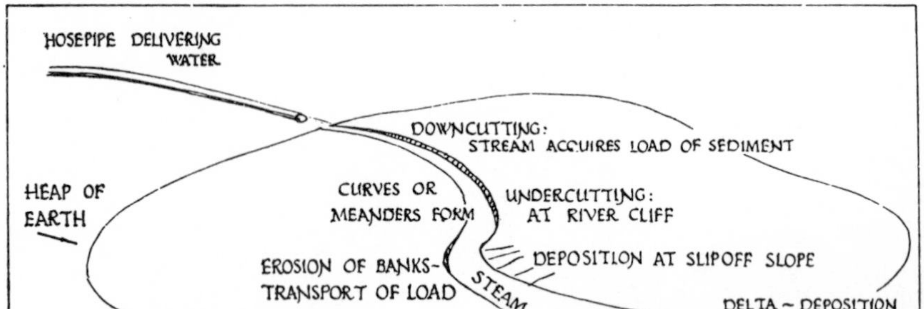

142

Measuring the speed of a river is an interesting task to give children when a river is crossed during a fieldwork excursion. A member of the party should be stationed at a point 'A' on a reasonably straight stretch of the river. He should throw into midstream a small piece of wood with some paper or wool attached. At the instant the wood alights on the water another pupil, armed with a watch having a second hand, should begin to walk along the bank from the point 'A' timing the progress of the float as he follows it. When the float has been moving for two minutes, the boy following it halts and then measures the distance back to the point 'A' (by pacing or any other method).

The number of feet that the float and therefore the current moves in two minutes is thus found. To find the speed of the current in feet per hour, multiply the result by thirty. The speed in miles per hour may then be found by dividing the number of feet per hour by 5,280.

The example of a river study given here demonstrates the wealth of material which is available in a well-chosen river study, and suitable too for a wide age range.

There is a very useful, but simple method that schoolchildren can use for plotting the long-profile and the cross-profile of a stream described in 'Geography in and out of School' by E. W. Briault and D. W. Shave.

An introduction to the quantitative aspects of stream study is valuable for the precision it brings and for the need it creates to use the methodology of other disciplines. Moreover, it is very illuminating for advanced pupils to realise that a river is not such a random process as it may at first appear. A study of a river meander is a case in point. In order to study a meander effectively certain terms need to be defined and discussed prior to the fieldwork, most of which can be easily understood. These are: width of channel (W), wavelength (Wv), length of channel (L), radius of curvature (R), point of inflection (i.e. the point where the curvature of the channel changes direction). It is important to realise that these elements of a meander are related by constant ratios e.g. Wv : R :: 5 : 1 (the ratio is 3 : 1 for a tight meander). L : Wv :: 1.3 The average ratio for the former is 4.7 : 1. The value of the latter ratio can vary between 1.3 : 1 and 4 : 1. These values reflect the sinuosity or tightness of the bend.

Measurements of these meander components should be made in the field. Later, ratios can be calculated to find out how far, if at all, your meander corresponds to the values given here. Other meanders should be studied for comparative purposes. The teacher may wish to test out two further river 'laws'.

(a) That the distance any river is straight does not exceed ten times its width at that point.

(b) That rivers contain deeps (pools) and shallows (riffles). Such riffles and pools occur at regular intervals along the channel, the spacing being related to the width. The riffles are located at intervals equal to about five to seven times the channel width, or roughly twice the wave-length of a typical meander.

This quantitative approach would strongly support other investigations, such as morphological characteristics (edge of terraces, river cliffs, slip-off slopes, braided channels, etc.) shown by landform mapping; hydrological features (variations in depth, velocity, direction of currents, bankfull stage etc.); and geological aspects (character of river bed, bank-soil section, and detailed mapping of edge of alluvium, sands and

143

gravels, areas of standing water, etc.). Ideally, the teacher should choose a meander which has not been altered or subject to human interference physically.

Also the study of a river can be approached from a more general point of view which has particular value for those not wishing to pursue the refinements of geomorphology to great depths. (Although the latter difficulty can be overcome by reading reliable texts on the subject, by consulting fellow members of the local branch of the Geographical Association, followed by a careful investigation alongside the local river.) This second method emphasises the relationship of human settlement and land use to a river valley. It involves tracing off the course (or part of it) of a river using the 6″ to 1 mile O.S. map to make a base map. The field patterns and settlement alongside the river should also be included.

The method is then to make a traverse along the river, recording not only recognisable physical features such as flood-plain, river terraces and the like but also vegetation, land use and cultural features in the landscape which are associated with the river. Field sketches and photographs can be added to the traverse diagram when it is completed to make a classroom display.

A Study of the River Wey near Guildford

1 *Aims* (a) The study of the physical and economic geography of a short stretch of the river in the Wey Gap. (b) The practice of field sketching and elementary surveying.

2 *Suitability:* this fieldwork may be undertaken by students at all stages of Secondary education.

3 *Time:* three and a half hours including lunch break.

4 *Distance:* two miles.

5 *Equipment:* 1″ Geological map sheet 269 (Aldershot). 2½″ O.S. map sheet SU 94. 6″ O.S. map sheet: Surrey XXXI NE old edition, or SU 94 NE TQ04 NW new edition. Foolscap, hardboard, clip, paper, pencils and drawing block.
The following might be useful: geological hammer, camera, surveyor's chain and range poles, binoculars.

6 *Preparation*

(a) *by the teacher:* the teacher should have some knowledge of the rock succession and structure of the Weald. He should be familiar with the action of rivers in eroding, transporting and depositing material. He should prepare a base map and walk over the route before taking his students. He may prepare a question sheet for pupils to answer either en route or on their return. With advanced students he should have read about the theories of the origin of Wealden drainage.

(b) *by the pupils:* with eleven-year-old pupils the features to be studied make so strong an impression that the most essential preparation is to study the route to be followed on the relevant O.S. maps. With more advanced students a base map should be prepared by each member of the class from the O.S. maps. The base map should cover the area shown on the sample.

144

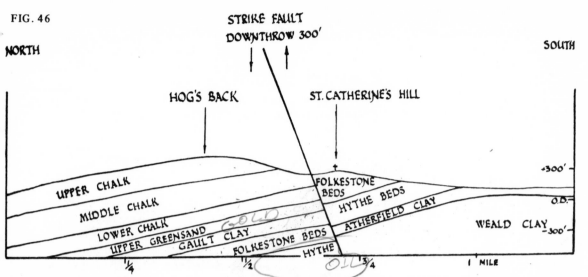

FIG. 46

STRIKE FAULT
DOWNTHROW 300'

NORTH

SOUTH

HOG'S BACK

ST. CATHERINE'S HILL

UPPER CHALK

MIDDLE CHALK

LOWER CHALK

UPPER GREENSAND
GAULT CLAY

FOLKESTONE BEDS
HYTHE

FOLKESTONE
BEDS

HYTHE BEDS

ATHERFIELD CLAY

WEALD CLAY

GOLD

OIL

+300'

O.D.

-300'

1/4

1/2

3/4

1 MILE

SKETCH SECTION THROUGH THE HOG'S BACK AND ST. CATHERINE'S HILL SHOWING THE EFFECT OF STRIKE
FAULTING

Sixth form pupils should have consulted a standard geomorphology on river work
and have digested relevant parts of *The Weald*, by Wooldridge and Goldring; *Structure,
Surface and Drainage in S.E. England*, by Wooldridge and Linton, *The Wealden
District (Geological Survey Regional Handbook)* and the *Aldershot Geological Memoir*.

7 *Route:* vehicles or coaches conveying fieldworkers to Guildford may be parked in
Millmead at 996492. Proceed southwards down Quarry Street. To your left rises the
Upper Chalk, to the right is the Wey flood-plain with Merrow Down and the Hog's
Back beyond. Turn right across the footbridge over the River Wey (997489).

What to note: the river meanders or twists and turns as it flows across its flood-plain.
Stand on the footbridge and face south. To your left is the river cliff where the river is
undercutting the bank and the road (shored up with concrete). Here the river runs
deep and fast, scouring the bank. To your right is the slip-off slope which is much
more gently sloping. Here deposition occurs where the river flows slowly on the inside
of the bend or meander. The course of the river is tending to migrate to your left at
this point.

Activity:
(a) compare as always the base map with the ground.

(b) Make an experiment to discover where the current of the river is fastest. Line up
the party across the footbridge in single file facing upstream. Give each a match; at a
signal their matches should be simultaneously dropped. After dropping their matches
the party should turn round and face downstream. The location of the fastest current
will be revealed by the first matches to emerge from beneath the bridge. These will be
fairly close to the right bank, and it will be readily seen how this bank is undercut,
forming a river cliff.

145

(c) Consider why the Quarry Street–Shalford Road runs where it does and the advantageous site of the quarry in relation to the river for transporting chalk and lime in early times.

(d) Sketch the river cliff and slip-off slope.

8   *Route:* cross the bridge, turn left along the river bank and proceed along the tow path to the sluice gate (996488).

What to note: on the far side of the river the water is kept from flowing over the flood-plain by a dyke or levée. The river water-level is above the level of the water meadows.

When the stream floods, some of the material or load carried by the water in suspension will be dripped close to the bank and this helps to form the raised bank or levée. In the case of the River Wey the levées have been strengthened and provided with sluices to regulate flooding of the river. But for these sluices the frequent flooding of the river would ensure that the levées would be broken open at intervals. Looking northwards you can see a good example of a river terrace on the right. The large sluice, constructed by a Fen country manufacturer, controls the water flow for the Thames Conservancy.

Activity:
(a) Questions to be answered:

What are the main features of the flood-plain?
What is it used for and why?
How is surplus water drained from it?
How would you recognise a river terrace?
How might the terrace have developed?

(b) Sketch the levée, its back-slope marsh, the flood-plain and a river terrace.

(c) Try to measure the speed of the current in miles per hour.

9 *Route:* proceed along the tow path to St Catherine's Ferry.

What to note: the damp flood-plain has many reeds. The slightly drier levée has only grass and rush. The back-slope marsh is covered with reed and scrub. The Folkestone Sands exposure on St Catherine's Hill will eventually be seen ahead of you. At 994485 a fault crosses the river at right angles to it and passes along the northern edge of St Catherine's Hill. There has been a down fall of rock on the north side of the fault and the Middle Chalk is brought directly up against the Folkestone Sands on the surface: the usual rock succession between these two beds, the Lower Chalk, Upper Greensand and Gault Clay have apparently been faulted out and are only found deep below the ground level (see Figure 47 p.148). The narrow valley along the line of the fault, at right angles to the river and parallel to the rock sequence, is an example of a strike valley. To the east of the tow path the fault passes beneath the alluvium of the flood-plain which was laid down by the river after the fault in the solid rock and appeared.

146

Activity:

(a) Consider why there is a ferry at St Catherine's and why it would have been important in mediaeval times.

(b) Study the spring which emerges from the base of St Catherine's Hill and draw a diagram of the River Wey's shortest tributary. Use terms such as source, tributary and confluence to label the diagram.

(c) With the aid of a geological map comment on the location of the valley between Merrow Down and St Catherine's Hill.

10 *Route:* climb St Catherine's Hill. There is a path leading eastwards from the river at the ferry which provides an easy route to the hill top, but the riverside cliff provides a more interesting ascent.

What to note: the river has undercut on the outer side of its meander. The bands of hard, dark brown ironstone are examples of carstone, bands of which occur irregularly in Folkestone Sands. The presence of the stone helps to explain the existence of the hill. The combination of porous, pervious beds of sandstone and hard layers of carstone resists erosion. The irregular layers of carstone follow lines of current bedding and are not examples of rock folding. The hill summit (994483) is a useful vantage-point for viewing the river features of the Wey Gap. The river meanders can clearly be seen on the flood-plain below.

The view to the north

Face north and orientate your geological map. You are now standing on the scarp crest of the Folkestone Sands. Below you, in the gulley runs the strike fault from east to west along the Vale of Holmesdale. The land on the north side of the fault has been thrown down about 300 feet relative to that on the south side, bringing the Middle Chalk directly against the Folkestone Sands (q.v.). The two resistant beds thus restricted development of a real scarp foot vale between the chalk and sandstone such as is to be seen in the distance on the eastern side of the Wey flood-plain. This is because no softer rocks are exposed at the surface (see Figure 46).

The main London–Portsmouth railway line, after tunnelling through the chalk, enters a tunnel under St Catherine's Hill. Tunnels were cut to avoid the built-up areas of Guildford and the areas liable to flood on the flat valley floor. The exposure of chalk visible in the face of the Great Quarry shows the dip of the chalk on the eastern side of the Wey Gap. The geological map indicates that the dip is greater on the western side.

The view to the east

Orientate your geological map. Pewley Hill, above the chalkquarries on the left, is the westernmost summit of that part of the Chalk Downs. Between this and the Folkestone Sands scarp of the Chantry Hills is a narrow vale – the Vale of Holmesdale. Having run under the river alluvium the strike fault is again exerting its influence and, for some distance along the vale, the Gault Clay and Upper Greensand are again

FIG. 47   BASE MAP OF GEOLOGY AND RIVER FEATURES IN THE WEY GAP

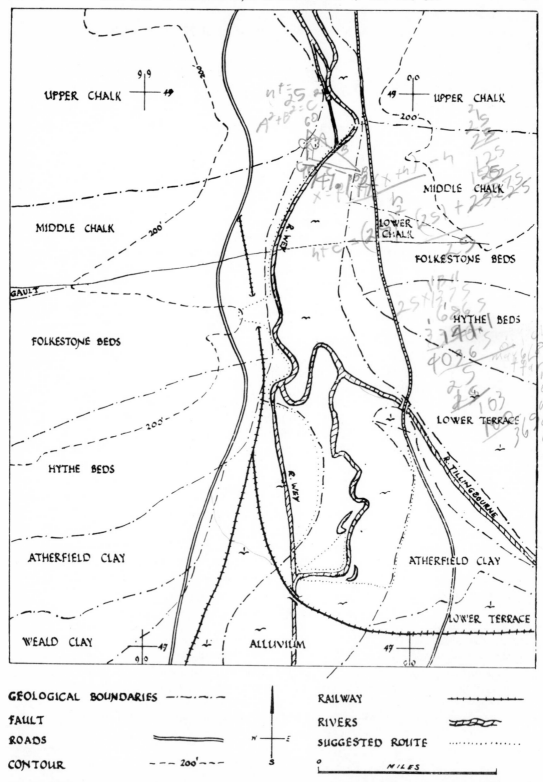

GEOLOGICAL BOUNDARIES — — — · — · —   RAILWAY   ┼┼┼┼┼┼┼┼

FAULT   ~~~~~~~~~~   RIVERS   ⟿⟿⟿

ROADS   ⟨═══════⟩   SUGGESTED ROUTE   ·············

CONTOUR   — — 200' — —

0   MILES

absent. Farther east the fault disappears, the softer rocks reappear at the surface in the normal succession and Holmesdale widens a little. In the middle distance the Guildford—Horsham Road skirts the flood-plain and Shalford Church can be seen. It is built on an old river terrace. The river meanders across the flood-plain, and the back-slope marshes are evident.

The view to the south

This is across the eroded anticline of Peasemarsh. Notice how the river and its tributaries have cut down deeply into the rocks weakened by uplift.

Activity: draw a landscape sketch of the view to the east. Label the sketch clearly.

11 *Route:* descend to the tow path and proceed southwards.

What to note: some chalk can be seen in the levée beneath your feet — evidence of its artificiality here. The next meander is interesting. Clearly, the strength of the northward flowing current may soon cut across the neck of the meander and leave the loop cut off as an ox-bow lake. At 995480 two channels of the river appear to converge. The westernmost channel is part of the Wey Navigation extended to Godalming in 1760. This consists of sections of canal cutting off some of the meanders which, however, still exist.

12 *Route:* at 996477 cross the navigation by means of the lock. Walk out to the oak trees at the edge of the terrace.

What to note: the lock is necessary to enable the navigation channel to surmount the river terrace. Away to the east the river course may be recognised by its high levées standing about the flood-plain. The river terrace near the lock consists of alluvium laid down by the river before it flowed quite as near to sea-level as it does at present.

Activity:
(a) study the lock to discover how it works.

(b) Sketch and photograph terrace and flood-plain features — levée, back-slope march and back-slope (sloping away from the river).

(c) This is a good place to measure the width of the Wey beyond the edge of the terrace. The width may be quite easily estimated by the following method. Take an object X such as a tree on the opposite bank: start off at right angles to it from A, and pace, say, ninety yards along the bank; on arriving at sixty yards plant a stick or stone, B; on arriving at C, thirty yards beyond that, that is ninety from the start, turn at right angles and walk away from the river bank counting your paces until you bring the stick and the distant tree in line; the number of paces that you have taken from the bank CD will then give you half the distance across AX. (See Figure 48 over page.)

13 *Route:* move south along the east bank of the navigation to the lock-keeper's cottage.

What to note: the cottage stands on the dry edge of the river terrace. There are good examples of back-slope marshes behind the navigation levées.

FIG. 48

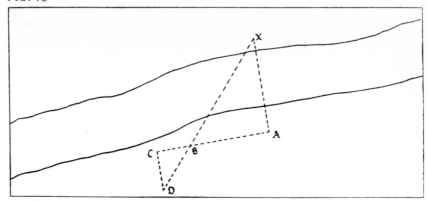

14 *Route:* from the cottage strike E.N.E. across the flood-plain to a prominent clump of trees on the west bank of the river (998475).

What to note: the abandoned meander has very little water in it but the old river bed can be recognised by its thick black mud. The higher ground within this old loop of the river is called a meander core, in this case it is thickly wooded.

Activity:
(a) consider the age of the trees in the ox-bowl and estimate when the river cut across the neck at the east end of the meander.

(b) Try to map this feature by the method of pacing triangles or by using the surveyor's chain and range poles.

15 *Route:* return to the lock-keeper's cottage and cross by the sluice gates. Turn left and walk along the river bank until you reach the clump of trees where the river turns north (998473).

What to note: the feature in the trees is another cut-off meander. The right bank is very steep where the river cuts through the neck of its old loop.

Activity: sketch, map and photograph the feature.

15 *Route:* from the riverside strike up the hill in front to a track at the top, then turn left.

What to note: this interfluve or hill between the Wey and Tillingbourne Valleys is made of alluvium and gravel.

Activity: review the river features of the flood-plain from your vantage-point.

17 *Route:* where the path rises steeply (999476) turn right to meet the Guildford—Horsham road.

What to note: in Shalford Village there are buttressed houses on the clay soil. The church is located on a terrace of dry alluvium and gravel capping the Atherfield Clay.

150

18 *Route:* continue northwards along the main road and descend to Tillingbourne Bridge.

What to note: the map shows that the Tillingbourne runs parallel to the Wey for some distance before joining it – a deferred junction is a common feature where rivers display characteristics of old age. An abandoned meander may be seen in the Tillingbourne Valley upstream of the road.

19 *Route:* cross the Tillingbourne and after passing the house on the corner turn left and follow the path on to Shalford Park. The playing fields may be on an old river terrace of the Wey.

20 *Route:* return to the main road and walk towards Guildford.

What to note: the road coming down from the chalk (998483) called Pilgrim's Way, and the footpath across the meadows to St Catherine's Ford both lend weight to the legend that this route was used by mediaeval pilgrims.

Comment: public conveniences are available on the playing fields and a convenient point for coaches to wait is at the entrance to Pilgrim's Way Road. If you continue walking towards Guildford you reach the footbridge where you commenced your study of river features of the Wey Gap. You have completed a circular tour of some of the best and most accessible examples of river work in the south of England. Why go to the Mississippi?

20

Fieldwork by the Sea

The Value of Coastal Fieldwork

Most geomorphological processes occur very slowly. Children find the slow movement of a glacier or the infinitesimally small uplift of a continental area very hard to appreciate. Coastal landforms, however, are unique because each day brings changes, often very dramatic and usually rapid. Such changes are easily observable and can be recorded. The greatest difficulty and potential danger is the shortness of time for study due to tides. The first vital requirement, therefore, is that the teacher knows his coastline thoroughly, that he has consulted tide tables and weather forecast and that he has left information about fieldwork location and plans with colleagues elsewhere.

The coastline has its own special appeal to children because it is a natural adventure playground. Pebbles on the beach; fossils in the cliff-face; rock pools; caves; lighthouses; quaint harbours full of cobbles and keel boats; the romance of smuggling — all these, and more, provide the teacher with a good starting point for fieldwork. Extra interest may come from dramatic events such as flooding; or from the excitement of lost villages, skindiving, submarine canyons and North Sea gas. Before embarking on fieldwork the teacher should try to develop this background in the classroom.

The coast combines opportunities for fieldwork in both physical and human geography. Often the latter will greatly influence the former in terms of sea wall constructions, groynes, piers, harbours, etc. But, ultimately the coastline integrates experiences better than many other locales. That area between cliff and sea is a distinctive environment in itself and to understand it effectively calls for a combined knowledge of geology, biology, geography, history and many other disciplines. Consequently, the elements and processes involved are often highly complex: the movement of shingle, the grading of beaches, the development of shoreline curves and cliff profiles, the action of the waves, are not fully understood, even by researchers.

The Scope of Coastal Studies

At first, coastal geography may seem a terrifying synthesis of all aspects of geography to both teacher and pupil alike. It seems to contain all that there is. The pattern and formation of dunes may require a knowledge of the physics of blown sand; the development of a headland such as Spurn Point may need the use of historical documents and maps; the study of a salt marsh calls for biological skill; the geological composition of cliffs, the trend of wave cut platforms and erosion levels may be hard to determine; coastal erosion and defence may seen to be more the realm of the civil engineer than the geographer; preservation of the coastline, access, the development of recreational facilities and the tourist industry may all seem to be part of the planners role. And so it goes on: justifiably the study of ports, estuaries and industrial coastlines might be included. What then is there for children to study by the seaside?

Fieldwork Activities

Related to the Physical Background

1   The Beach

(a) *Mapping the beach: beach plans*

Aims: to understand beach form and structure. To develop keen observation of minor landforms. To practice mapping techniques.

Activities: i. Prepare a zig-zag book in class to show relief, geology, drainage, land use, settlement and communications for a selected coastal strip (i.e. the field area). Each element on a separate page. This is an individual exercise to give background to fieldwork. Sections, annotations, etc. should be added. Resort guide books can be very useful. (2½" O.S. map to be used if possible).
ii. Prepare a diagram which shows typical beach features and nomenclature. Duplicate copies and discuss in class. This base diagram can later be used as a field exercise in which the children have an individual copy to which they add details as observed, particularly noting typical features. (Figure 49a, see over).
iii. Using a 6", 25" plan and/or the relevant chart for a small part of the coastline make a base map. In the field add beach details and also try to make a simple morphological map which will indicate such items as cliff landslips, fault-lines, coastal ravines, wave cut platforms.
iv. Select a suitable part of the beach which displays several contrasting zones. Find Grid reference. Walk from base of cliff towards the sea. Sketch-map the area traversed in field books pacing or measuring each zone and indicating its character by agreed symbols with annotations. Later turn the rough field map into a beach plan like Figure 50 on page 155. Relate to a simple beach profile. Annotated photographs and small pebbles can also be added to make a full traverse/transect display.

A series of beach plans should, if possible, be made and later compared for several beaches separated by, say, 200 yards. The magazine *Which* did an interesting survey of resort beaches in the Isle of Wight to find out which would appeal most to the holidaymaker. They used beach plans to bring out the main features.

Since beaches may change rapidly it would be useful to make sample maps at different times of the year if possible. Observed changes should give scope for class discussion on the effect of weather and sea conditions at different times of the year.

(b) *Beach measurements*
More accurate survey methods may be used by the teacher if he wishes. This would help to establish beach profiles and gradients more effectively (see C. A. M. King, *Techniques in Geomorphology*). Width and gradient of various beach zones can be

153

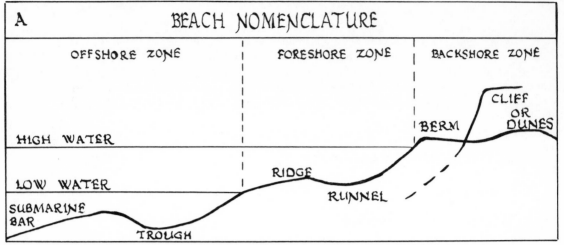

A BEACH NOMENCLATURE

OFFSHORE ZONE · FORESHORE ZONE · BACKSHORE ZONE

CLIFF OR DUNES

BERM

HIGH WATER

RIDGE

LOW WATER

RUNNEL

SUBMARINE BAR

TROUGH

FIG. 49A

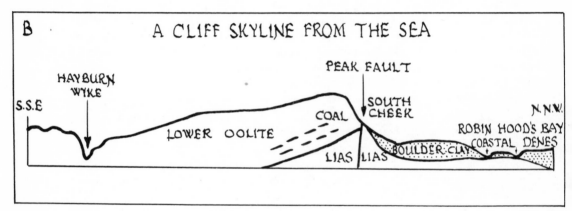

B A CLIFF SKYLINE FROM THE SEA

PEAK FAULT

HAYBURN WYKE

S.S.E

COAL

SOUTH CHEEK

N.N.W.

LOWER OOLITE

ROBIN HOOD'S BAY
COASTAL DENES

LIAS LIAS

BOULDER CLAY

FIG. 49B

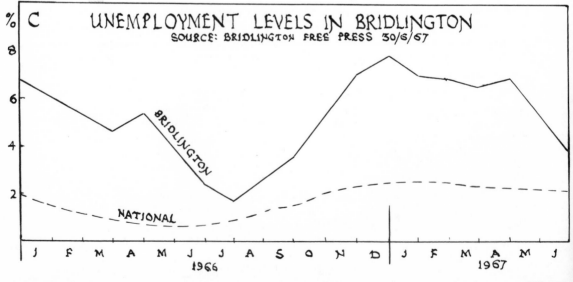

C UNEMPLOYMENT LEVELS IN BRIDLINGTON

SOURCE: BRIDLINGTON FREE PRESS 30/6/67

%

8

6

BRIDLINGTON

4

2

NATIONAL

J F M A M J J A S O N D J F M A M J
1966   1967

FIG. 49C

154

SEA LEVEL

1½' DROP

1' DROP

1' DROP

SEA

SAND WITH A PEBBLE RIDGE
LARGE PEBBLES AND BOULDERS
IN FRONT

38'

8'

ROCK PLATFORM WITH SAND DEPOSIT-
ION AND SEAWEED IN THE SMALL
ROCK POOLS

8'

20'

STEPPED & FLAKY ROCK PLATFORM
WITH OCCASIONAL BOULDERS, SAND
& GRAVEL CHANNELS AND PEBBLE
RIDGES

80'

DEPOSITION GRADED FROM SAND
TO PEBBLES & SMALL BOULDERS
AS IT GOES TOWARDS THE EDGE
OF THE ROCK PLATFORM.

36'

20'

ROCK PLATFORM COVERED IN
SEAWEED, BARNACLES & LIMPETS

25'

DEPOSᴺ. OF PEBBLES & SMALL BOULDERS

15'

ROCK PLATFORM WITH NO SEAWEED

25'

LARGE ROCK POOL WITH BOULDERS
COVERED IN SEAWEED

30'

PROFUSE SEAWEED OVER ROCK
PLATFORM. SNAILS & LIMPETS

45'

SEA

155

measured and compared to see if any meaningful relationships can be found, e.g. ratios in a river meander. The direction of wave incidence should be noted.

(c) *Pebbles on the beach*

Activities: i. Collection and identification of pebbles and fossils using C. Ellis, *Pebbles on the Beach* and Himus and Sweeting, *The Elements of Field Geology.*
ii. Special study of erratics: try to identify, suggest origin, discover most commonly found erratics along a stretch of coastline as a basis for later class work on ice movements. This can be done well along the Yorkshire coast (see Figure 2 in C.A.M. King, *The Scarborough District*).
iii. Using sieves with varying gauges try to establish pebble zones according to size. Map. Sampling of pebble masses, permeability of the beach material and other interesting methods are described by P. R. Corbyn, 'The Size and Shape of Pebbles on Chesil Beach' in *Geographical Journal*, March 1967 and Higgins and Parkin, 'Sea Waves and Beach Cusps', *Geographical Journal*, June 1962.
iv. For schools near the coast, make a geological rock garden at school using beach materials. Label well. Flotsam and jetsam can be included. The teacher should remember that the many and varied shapes, sizes, colours and patterns to be found on the seashore provide stimulating material for art work as well as geography.

2 The Cliff

Aims: to study cliff morphology and geology, in particular, the contrasts along a given length of coastline. To investigate natural and man-made erosion.

(a) *Cliff profiles*

Activities: i. Sketch cliff profiles at selected points. Add geological details later using memoirs, geological map, etc. Attach finished versions to a coastal map as in Figure 51. Relate geology to profile in detail to explain landslips, cave formation, Undercutting, etc.
ii. Sketch/photograph coastal forms, e.g. stacks, arches, blow-holes. In class make a detailed annotated diagram from a photo and try to show in a series of such line drawings how the stack evolved.

(b) *Cliff face and skyline studies*

Activities: i. Calculate cliff height at selected points.
ii. Draw a diagrammatic section of the cliff face. Idenfify geological horizons.
iii. Teacher to provide a cliff skyline as seen from the beach (*Admiralty Pilot Books* may help in this). Pupils to mark in visible features also using 2½" O.S. map in order to eventually relate geology, relief and drainage.
iv. List man-made features along the cliff, e.g. wartime defences, culverts, gravel workings, etc. Attempt to assess erosional effects of the sea and note erosion from the landward. Try to study how certain rocks have been more resistant than others, in particular note the different effects of erosion in producing distinctive landforms, e.g. pyramidal gullying in boulder clay.

156

FIG. 51

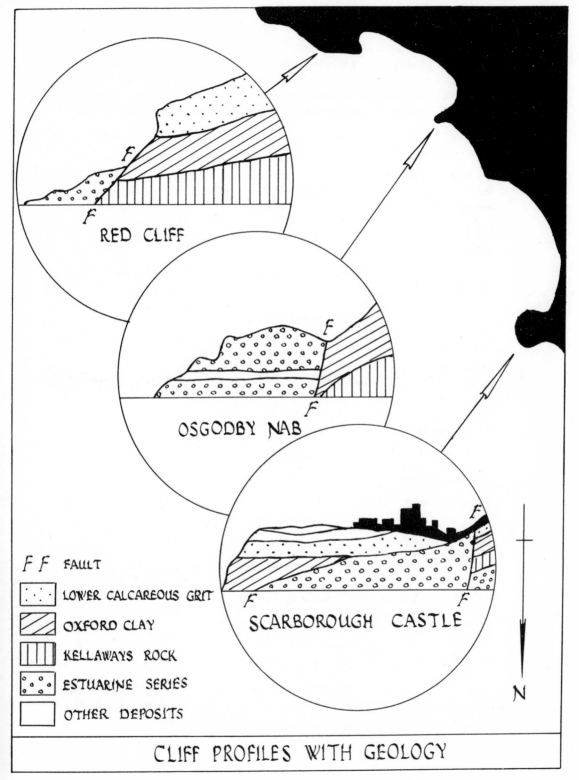

RED CLIFF

OSGODBY NAB

SCARBOROUGH CASTLE

F F FAULT
LOWER CALCAREOUS GRIT
OXFORD CLAY
KELLAWAYS ROCK
ESTUARINE SERIES
OTHER DEPOSITS

N

CLIFF PROFILES WITH GEOLOGY

3   Geography from the Sea

One of the best places to study the coast is from the sea, yet very few field parties attempt such work. It is true that there may be many practical difficulties but it is worth over-coming these to gain a unique and memorable educational experience. Besides, the cost per head is often relatively cheap in the off-season for a motorised coble with a capacity of about thirty-six. Many worthwhile activities are possible:

(a) Class to prepare a briefing on weather, sea conditions, visibility and tides for the cruise using radio broadcasts, newspapers, tide tables.

(b) Prepare a flash card detailing navigational features likely to be seen, e.g. buoys, channels, lights, etc. using *Admiralty Pilot Books, Manual of Seamanship, Nautical Almanacs, Observers Book of Ships.*

(c) Class study of relevant charts to understand symbols. Information on coastal conditions from *Pilot Books.*

(d) Class might study a headland from the land one day, and the next day do the same from the sea. They would be equipped with a duplicated map showing coastal details only. Their task would be to identify landmarks and to take compass fixes on them, noting them on the map. Cliff skylines obtained from *Pilot Books* and memoirs would be issued as the base for notes during the cruise (Figure 49b, p. 154). Sketching and photography would also be possible especially of folding in the cliff face. Later, the photos can be fully analysed and line drawings prepared. Release of small plastic containers at sea with instructions for return might yield some useful information about coastal currents.

4   Geography from a Lighthouse

(a) Viewfinder exercise using binoculars.

(b) Map reading.

(c) Ship spotting.

(d) Working of the lighthouse: collect and list regulations for safety at sea – visit lifeboat station. Siting of the lighthouse.

(e) Photography from the lighthouse: analyse later and compare to O.S. map.

Related to the Human and Economic Background

There are many interesting possibilities for fieldwork here.
Briefly they might include:

(a) A unit study of a bay, headland or estuary.

(b) An historical geography study of a castle or abbey on a prominent headland from the point of view of stragetic siting, building stones, etc.

(c) A study of fishing in a small village, e.g. Staithes.

158

(d) A detailed harbour study to contain: siting, approaches, physical growth, land use of piers, boating facilities, number and type of vessels, berthing charges, marina development, etc.

(e) North Sea gas: an investigation into the effects on the landscape and economic life of the region. Background details from PIB publications.

(f) A planning study to discover the development of subtopia along a particular coast. See W. T. Rees-Pryce, 'Location and Geography of Holiday Caravan Camps in Wales' in *Trans. Inst. of British Geographers*, December 1967. Note also *Enterprise Neptune*.

(g) The tourist industry: the study of holidaymakers. Since this is such a wide ranging and important item an actual example of how to set about making a study has been included.

How to Study Holidaymakers: The Tourist Industry in Whitby

The moors and the sea may hold potential riches, such as potash and natural gas, but the most important industry of the Yorkshire coast since the 19th century has been tourism. The magnificent scenery and the depth and richness of history along the coast are still the most significant resources of the region.

Scarborough was the earliest of the Yorkshire resorts, being much visited in the early and middle 17th century. Whitby did not achieve distinction until the 19th century. It was after 1848 that the improvement of rail communications and the rapid growth of nearby industrial areas such as Tees-side promoted the development of Whitby as a resort.

1   The Holidaymakers

These are residential visitors and day trippers. It is difficult to get information about the latter usually it has to come indirectly from a study of car/coach parks, day returns sold by the railway, and estimated passenger load on buses, trains and private cars. Sometimes a great deal of information can be gathered by means of the interview/questionnaire which can be conducted on the beach, in the street, or at one of the local entertainment centres. However, the questionnaire may be easier to operate with residential visitors since forms may be left at hotels, boarding-houses, the Spa, the Public Library and Museum, providing, of course, that permission has been previously obtained. In addition, street interviews should also be attempted at different times during the season. The following questionnaire was used most successfully in a survey in 1956:

Sex   Age   *Time   Date   Place*
1 Where are you from? Place, County.
2 What is your occupation, or that of the head of the family?
3 How many persons are there in your party? Adults, children.
4 How did you travel to Whitby?
5 How long are your staying in Whitby?
6 Is this your first visit to Whitby?

7 Why did you choose Whitby for your holiday?

8 Where did you go for your holiday last year, the year before?

9 What type of accommodation have you in Whitby?

10 What do you find attractive in Whitby?

11 Have you any criticisms of fine weather facilities in the town?

12 Can you suggest improvements to wet weather facilities in the town?

13 What trips have you made/do you intend to make whilst staying here?

14 Can you say approx. how large a budget you are allowing for total holiday expenditure this year?

15 Any further comments?

Analysis of findings will give a very full picture of the holidaymaker. It will be possible to draw maps/graphs/diagrams of origins, occupations, age structure, mode of travel, type of accommodation, length of stay, trips out, etc. These may be assembled into a display, or, better still, made into a class-book which can be revised every few years by other groups.

The value of such a questionnaire can be seen by the summary findings of the 1956 survey, some of which were:

(a) Holidaymakers are mainly from the north of England especially the large cities.

(b) The proportion of children is higher than the regional and national averages.

(c) Weekly visitors are most significant: those staying a fortnight are important in the peak period later July—early September.

(d) Accommodation in private houses rivals that of licensed hotels and boarding-houses. The latter having suffered a decline.

(e) The natural beauty of the area, the beach, the harbour, historical interest, sports and entertainments were respectively most popular.

2   Investigation into the Advertising of Whitby as a Resort

Find out from the Information Bureau exactly what advertising is done, when and where. Examine regional and national newspapers and magazines for references to Whitby (e.g. *Yorkshire Post, Dalesman*). Does the Post Office have a special franking, e.g. More sun, more sand at Sandown . Investigate hours of sunshine at selected resorts. This may be extended to rainfall, temperature, etc. Compare your resort with others elsewhere. Make a chart from this. If you can get details of postal enquiries made at the Information Bureau over a period you will be able to draw a map which helps to indicate where visitors are likely to come from. Is there a federation of the Yorkshire resorts for the purposes of advertising? Find out more about this from the Chamber of Trade. Examine National-Park publications, etc. to see the quantity and quality of the references to Whitby. How far does the advertising seek to create a particular impression of Whitby? What is the impression? What is the actual expenditure on advertising? When are the 'Wakes Weeks' in the industrial cities? Is the advertising aimed in this direction? How much private advertising goes on by individual hotels, etc.

3 Investigation into the Food supplies for the Visitor

Enquiries at local dairies, abattoir, market etc. will give a good idea of the origin of milk, meat and vegetables. How much comes from local farms? Plot these on a map. Pinpoint also nearby poultry farms, glasshouse crops, potato-growing areas and fruit growing within a few miles of the town. Try to find out other sources of supply, e.g. tomatoes from the Hull area.

What proportion of fish landed is actually bought for use in the town? Visit fish auctions, Whitby Fish Selling Co., Harbourmaster.

4 How has the Holidaymaker changed over the Years?

Analyse old and new town guides, Ward Lock guides, to discover the changing character of amenities offered, noting differences in appeal, tone, clientele etc.

5 How do the Holidaymakers arrive?

The questionnaire will help considerably to answer this question. However, there are a few other ways of findings out:

(a) *Car park analysis:* The registration number of a vehicle indicates its origin. Thus, in 545DYN the YN is a London registration. The *A.A. Handbook* gives the code for all registration plates. It is possible, then, to collect registration numbers of cars in a specified parking area and later to plot origins on a map. Once this has been done it then becomes clear where car drivers came from on a particular day. Can you think of the defects of this method?

Choose the central area of the town (define this clearly) and collect registration numbers of all cars in car parks, streets, etc. Note time/date carefully. Plot results on a map of the British Isles showing county boundaries. Several surveys should be taken during the holiday season and in winter months.

(b) *Coach parking analysis:* coaches have owner and origin marked on them. Visit coach parks, etc. noting geographical origin/date/time. Try to find out if these are day visits. Plot the results on a map. Note maximum journeys and duration of stay in the town is possible.

6 How well is Whitby served in winter and summer by rail/bus services?

Analyse services from available timetables producing a table similar to the following:

Number of Trains per week to & from	Miles by rail	Time of journey	Cost

Make a comparative table to show rail and bus services. Can you draw a frequency diagram for winter and summer? Note express coach services. Would it be quicker to go to Glasgow by coach or train?

Assess the effect of the axing of local railways on the total development of the resort. Does it mean the end of the line for Whitby? Make a survey in your school to find out how many pupils are affected; Use *'The Reshaping of British Railways'* (The Beeching Report) and the local newspaper reports to help your enquiry.

It may be possible to obtain information about the number of tickets collected at Whitby Town station and West Cliff. These can be tabulated.

The siting of the United bus station next to the main railway station seems ideal. Examine the quality of both sites in relation to hotels, etc. What facilities exist for the movement of visitors' luggage from both stations?

To what degree would you say that Whitby is difficult to reach by car, rail, bus? Study the O.S. map noting steep gradients, road patterns, direct and indirect routes influenced by valleys, etc. Compare with Scarborough and Bridlington.

7   A study of Holiday Amenities

(a) List fine weather facilities, distinguishing natural and man-made ones. Plot the major facilities on a map. Include such details as street lighting, conveniences, free seating etc.

(b) List wet weather facilities. Do you consider them satisfactory? Find out the usage if possible. When does the season begin and end? How can you determine this? List star attractions, festivals, etc. What efforts are made to prolong the season? (at Scarborough cheaper holidays for pensioners, the Cricket Festival, Dutch Week, etc.)

(c) Make a special study of a section of the beach looking at it as an amenity. Define your area first, locate it re. town; access, paths, roads, steps, etc. How frequently can it be used? (Study tide-tables, note exposure to winds etc) Examine the quality of the environment around: is it semi-rural, built-up, surrounded by amusements? Make a rough count of people using this section of beach at selected times: note age/sex composition. Try to find out why they chose this part of the beach and if they usually go back there. Make a map of the beach showing the distribution of sand/stones/rock pools. HWM LWM, cliff line, and mark the position of the nearest ice-cream kiosk, tea stall, telephone, first-aid hut, deck-chair store, etc. Particularly note if they are absent. Is there any provision for sea rescue?

(d) Make a survey of the central areas around Whitby Bridge using a base map and mark on it restaurants, cafes, snack-bars, public houses, fish and chip shops. Analyse the location pattern, in particular notice if concentrations decrease outwards.

(e) Make a study of the harbour as a tourist attraction. List what you consider to be most attractive. Consult the harbourmaster and Information Bureau to discover numbers of rowing boats, pleasure craft available. A good skyline sketch can be drawn from West Pier Lighthouse giving an idea of the functional and historical growth of the town. This can be compared to earlier prints, engravings and sketches.

(f) Examine the quality of the coast north and south of the town (fieldwork, 1″ and 2½″ O.S. maps) in terms of an amenity. How accessible is it? (Consult bus/train timetables.) Draw a coastline map/plan to show its contrasting appeal − insert cliff heights, valleys, rock ledges, sand areas, etc.

162

(g) List inland attractions and give them a star-rating according to accessibility, scenic value, etc.

(h) Examine freedom of movement in central Whitby. Take a traffic census at the Bridge for different dates and times   note percentage of through traffic. How frequently is traffic interrupted by the swing bridge? To what degree do the approaches rather than the bridge cause congestion? Investigate the approaches on both sides (photograph congestion). On a map of the central area mark parking spaces, street parking, one-way streets. Select one street and over a period measure pedestrian-flow along pavements. It may be possible to obtain details of accident-locations for several years from the police Annual Report. These can be added to the map.

(i) Make a study of the coming of the railway to Whitby, especially the opening of the different lines. Pay particular attention to George Hudson, the 'Railway King', local opinion, and the effect of the railway on the growth of the town (e.g. West Cliff estate).

An excellent fieldwork topic would be a railway traverse along part of the Whitby–Pickering and/or the Whitby–Scarborough line, especially recording and evaluating the scenic quality (see H. Belcher, *Scenery of the Whitby and Pickering Railway*, 1836. The Ravenscar and Newtondale sections are outstanding. Examine the physical background and constructional difficulties (use of $1''$ $2\frac{1}{2}''$, and geology maps here). Note tunnels, embankments, gradients, adaptation to local relief, special problems, e.g. Fen Bogs peat area, land use along the lines, especially industrial uses such as alum and iron working, evidence of mineral lines. Photograph and sketch from viewpoints, collect soil samples, etc. Some special detailed studies might be attempted, e.g. a deserted railway station; 'the town that never was' at Ravenscar.

Assemble work as a traverse/transect chart in the classroom and a full study of the railway as a whole can then be made. This can lead on to a wider examination of railway geography.

(j) A shop window analysis: Select a typical tourist shop selling items such as jet or pseudo-jet goods, shell ornaments, sea-urchins, etc. Make notes describing the site of the shop, pedestrian-flow, antiquity of the building, etc. Then draw a diagrammatic sketch of the layout of the window highlighting items actually made in or near Whitby. Find out by enquiry inside what is most in demand. A front elevation of the row of shops can be drawn to show age, building materials, usage of different floors. (Analysis of a wet fish shop can be similarly done to show the range/type/origin of fish available.)

8   "After the Ball is Over . . ."

Comparing a resort in summer and winter is like comparing the living and the dead. A resort is, in reality, two very different towns. How can you examine this more closely?

(a) Unemployment: basic figures are available from the local employment exchange. They are also given in the newspaper. Try to graph them as in Figure 49c on page 154.

Find out which age groups and occupations are most effected. What is the position of the fishermen? Some of the Yorkshire resorts are in the Development Area (Whitby,

163

Scarborough, but not Bridlington). Find out exactly what this means — how beneficial is it to be included? Has it stimulated industrial expansion in Whitby? Where? What industries? What is the effect on employment?

(b) Plot on a map of the sea front and central areas the establishments which have closed down or limited their opening during winter months.

(c) Using bus/train timetable deduce the restrictions in services for winter. Collect examples of weather interfering with transport. Further analysis of car parks will emphasise seasonal contrast.

If you wish, you can extend your fieldwork to include accommodation studies in which an hotel study might be attempted.

21

A Study of Coastal Features

For many schools a coastal study involves a fair amount of time spent in travel before the coast is reached. Therefore such work is often best left to a week's school journey to another part of the British Isles rather than attempted as a separate day's excursion.

The purpose of such a study is perhaps most important if it sets out to give the pupil an understanding of how the coastline is altered by the erosive and depositional effects of the sea. It has added value too if it is experienced as a contrast to fieldwork in the home area. The following study in the Hythe-Folkestone area of Kent is an example of the wealth of fieldwork that can be enjoyed in a coastal investigation.

Aims: to study the constructive and destructive work of the sea in a small area, and to discover how human activities have modified and been modified by it.

Suitability: this exercise has been planned for children in secondary schools between the ages of 11 and 15. It could be adapted for use with 9- and 10-year-old primary school children.

Time: about four and a half hours in the locality. The journey time from the Purley area of Surrey takes three hours, including one stop.

Distance: the total walking distance is two miles, allowing for the use of the coach between Hythe and Folkestone.

Equipment for Teacher: sheet 12¼″ O.S. map, Sheets TR 13 and TR 23 2½″ O.S. maps, Sheet 305/306 1″ Geological map. The following would be useful: binoculars, camera, hammer, small blackboard and chalk.

Equipment for Pupils: each pupil should be issued with printed extracts from the ¼″ map (adapted according to age) for use during the coach journey. They should carry a piece of hardboard with a base map of Folkestone mounted on one side and a list of questions with spaces for answers on the other side. Every pupil must have pencils, erasers, mackintosh, walking shoes and packed lunch. Additional refreshments may be obtained en route as indicated in the itinerary.

Preparation by the Teacher: adequate quantities of the ¼″ route map, Folkestone base map and questionnaire should be duplicated. A circular should be sent to parents indicating time and cost — and details of these must be obtained from the coach company. It is an advantage to make this study when the tide is low, this can be worked out accurately by consulting Tide Tables, or approximately, by using the

FIG. 52

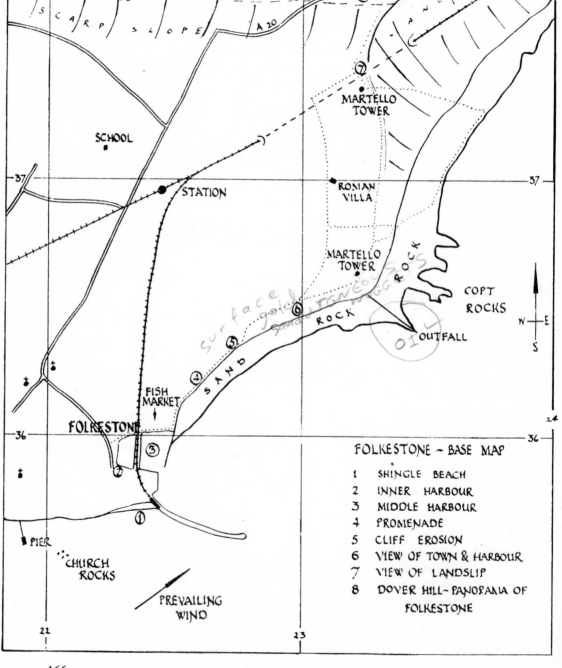

SCHOOL

"VALIANT SAILOR"

⑧

A 20

SCARP SLOPE

500'

⑦

MARTELLO
TOWER

LANDSLIP

TO DOVER

STATION

ROMAN
VILLA

MARTELLO
TOWER

COPT
ROCKS

Surface

gold ignéous

sand igneous rock

SAND

ROCK

⑥

⑤

OIL

OUTFALL

N

W   E

S

④

FISH
MARKET

FOLKESTONE

③

②

①

PIER

CHURCH
ROCKS

PREVAILING
WIND

FOLKESTONE — BASE MAP

1   SHINGLE BEACH
2   INNER HARBOUR
3   MIDDLE HARBOUR
4   PROMENADE
5   CLIFF EROSION
6   VIEW OF TOWN & HARBOUR
7   VIEW OF LANDSLIP
8   DOVER HILL — PANORAMA OF
        FOLKESTONE

tables in the *A.A. Handbook*. Familiarity with the geological sequence of the Weald is necessary and special attention should be paid to the four beds of the Lower Greensand. The main effects of the processes of marine erosion and deposition should also be known. A large-scale map showing drift geology could be produced for class display. Teachers interested in photography could print copies of a suitable photograph for children to annotate in the field. Census figures should be consulted and population returns for Folkestone and Rye could be compared. The sudden build-up of Folkestone's population during the last century was obviously connected with the coming of the railway and the extension of harbour facilities for the cross-Channel service.

Preparation for Children: using an atlas and ¼" map they should study the overall position of Folkestone in relation to the Channel, France, the North Downs and the route to London. They should be briefed about the coach route to Folkestone and the work they will be expected to do both in the field and afterwards. They should all know the geological cross-section from the North Downs to the Weald Clay Vale. No introduction to the work of the sea is essential but an elementary understanding of wave action and knowledge of longshore drift would be useful.

1   Coach Travel to Hythe

Some simple task to be carried out during the journey is desirable so that the pupils' attentions are focused outside the coach. As the main route follows Holmesdale and goes through or past a number of gap towns, attention could be concentrated on these.

Activity:

(a) Use the ¼" route maps. These should already show the North Downs and major gaps, the Greensand Ridge and the main A 20 route which is the one to follow. Gap towns should be marked but not named, and at these towns important road junctions should be shown and the roads numbered, so that children can recognise the town, by watching the road signs, and mark the names on the map where indicated. Depending on where the A 20 is joined, the main settlements to be marked are Godstone, Sevenoaks, Maidstone and Ashford. The rivers should already be marked and named on the map.

(b) *Questions to be answered:* what are the common features of the towns marked on the map? What difference are there in farming and vegetation between the hills to the north and south and the vale in which you are travelling?

Comment: a convenient stopping-place for toilets and refreshments is the 'Roebuck' at Harrietsham between Maidstone and Charing.

2   Hythe Promenade

Arrival time 11.30 a.m. Park the coach at 'The Green' (158345) and walk to the corner of the promenade at 158339.

What to note: the dominant features here are the marine deposition, such as the shingle beach, the beach ridges related to normal high tides and storms and the groynes put up to control longshore drift. The curving shore-line of Dungeness can be seen to the south-west. This serves to stress how the land has grown at the expense of the sea, although the significance of this can best be seen at 'The Green'. Striking human features are the three old Martello Towers indicating the vulnerability of this flat shore-line and the defensive measures against invasion taken in Napoleonic times. The view of the nuclear power station on the skyline can be used to stress the siting of these buildings in isolated and infertile areas, and near a supply of water for cooling the reactors. The rifle ranges and fishing vessels should also be noted.

Activity:

(a) *Questions to be answered:* of what material is the beach made? Is there any evidence of the extent of high tide? Why does the beach rise in a series of steps? Why is there a series of wooden structures at right angles to the shore and do they have any effect on the shape of the beach? What was the purpose of the towers along the coast? What evidence is there of military activity? Where are the small fishing vessels registered?

(b) Attempt a sketch of the shore-line looking eastwards, marking in the present water mark, high-tide mark, shingle ridge, groynes and probable angle to the shore of the dominant waves.

Comment: as one looks westwards from the end of the promenade towards Dungeness the storm beach is very noticeable; in the opposite direction it is not a significant feature and is replaced by the promenade. If the beach is likely to be covered by the tide it would be best to study the Hythe section after the visit to Folkestone, especially if there is no intention to proceed to Dover.

3 Route

Return to the Green'.

Note: To the west the old shore-line is clearly seen from here. The relationship to this coast of the old Roman port and of the main part of Hythe can be referred to at this point.

Activity: questions to be answered: what would happen if the sea-level rose by about ten feet? Where would the coastline be? What sort of material is the flat area between the old coastline and the present shore made of? What do you notice about the situation of the main part of Hythe?

4 Route

Travel to Folkestone by coach following the A 261 and A 255 into Folkestone.

What to note: as you travel east from Hythe the flat coastal area gradually narrows and the scarp slope of the Greensand Ridge approaches the coast at Sandgate. Here the road leaves the coast and climbs to the main part of Folkestone which has developed on the dip-slope. On the left of the road are signs to Shorncliff Camp, the main army barracks of the town. The establishment of the army in the area in Napoleonic times as evidenced in the barracks, Martello Towers and Royal Military Canal, led to the growth of Folkestone as a resort. Army families settled here, especially when soldiers were abroad for long periods. The further expansion of Folkestone was related to railway development and will be referred to later.

Activity: list evidence of military activity. E.m.O. — camburry Geses f. 6 02

List evidence which indicates that Folkestone is a resort.

5 Route

Proceed by coach around the west side of the harbour to the coach park (231357). The time of arrival here should be about 12.15 p.m. Before leaving the coach, arrangements should be made with the driver as to where and when the party should be picked up, then proceed through the amusement park on to the beach.

What to note: there is a considerable accumulation of shingle behind the harbour pier which was built in 1861 and extended in the 1880s. The sudden development of a beach contrasts with the cliff-lined shore at Sandgate. The main reasons for the recent accumulation are the seaward extension of the pier and the direction of the longshore drift.

Activity:

(a) *Questions to be answered:* which way is the wind blowing? What process is causing the accumulation of shingle? Where is the shingle being removed from? Why is the accumulation on the western side of the pier?

(b) Attempt a sketch map to show where erosion and deposition are taking place, the pier and the old shore-line. The base map could be used or the teacher may prefer to draw the map on a small blackboard brought for the purpose.

(c) This is a good point for taking photographs.

6 Route

Proceed from the beach to the promenade and walk to the inner harbour near the level crossing.

What to note: from here there is a good view of the inner harbour, the railway and its steep incline and the upper part of the town. Ferry timetables may be studied. It is a convenient point to stress Folkestone's function as a cross-Channel port. One reason for the restricted ferry service is the difficult railway approach. The gradient is very steep although the line uses the valley of the little stream. At its mouth this stream has been diverted so that its waters scour the inner harbour. Compared with Dover, Folkestone has always been a minor port and, although a member of the Confederation of Cinque Ports, she was not one of the seven Head Ports. The railway connection to London, established in 1843, has been responsible for the expansion of both port and town.

It is possible that hotel accommodation is connected with the cross-Channel passenger service as well as meeting holiday resort and residential requirements.

Activity:

(a) Study the ferry timetable to find out the main use of the harbour.

(b) Suggest any obvious advantage of the position of Folkestone as a cross-Channel port.

(c) Suggest any obvious disadvantages of the site.

7 Route

Walk around the harbour and under the railway viaduct to the Fish Market. On the way note the profile of the Greensand Ridge.

What to note: there is a good view of the inner, middle and outer harbour and the activities associated with each. Another limitation of the port can be seen in the exposure of the outer harbour to easterly winds. The close proximity of many inns could be used to give a simple lesson in economics.

Activity: questions to be answered: What types of fish are on sale? What varieties of shellfish can you identify? How many fishing boats are in the harbour and how big are they? What other evidence is there to show that Folkestone is a fishing port? For what other purposes are the three ports of the harbour used?

8 Route

Proceed to the end of the Middle Harbour.

Activity: many ships should be visible on the horizon and children should find out how far they can see. The rough formula for discovering the distance to the sea-level horizon is to find one's own head height, in feet, above sea-level, add 50 per cent and find the square root. The answer gives the distance seen in miles.

9 Route

Proceed to the end of the new promenade (227364).

What to note: in great contrast to the beach west of the harbour there is a marked change in the shape of the coastline to the east. Here the group should be able to reconsider the development of the coastline as this is a good place to observe the very obvious onset of coastal erosion. There is a sandy beach, as compared with shingle to the west, and a concrete wall and promenade have been built to prevent further attack by the sea. The capping of blue-grey Gault Clay which overlies the Folkestone Sands can be seen slumping down the cliff face, and farther east there are good examples of cliff falls. At high tide the sea washes under the promenade and removes any clay that may have fallen. It is able to do this because of the form taken by the promenade. Differential wind erosion and possibly spray from storm waves etch the varying bands of the Folkestone Sands revealing narrow bands of limestone and harder slabs of sandstone.

170

Activity:

(a) *Questions to answer:* what do you notice about the shape of the coastline here as compared with that near the coach park? Is the sea adding to or taking away from the land here?

(b) Draw a diagram to show how the promenade is constructed and comment on its suitability as a form of sea defence.

(c) As you walk along the promenade find as many pieces of evidence as you can to show that the land is being worn away.

Comment: The following show that erosion is occurring. Blue Gault Clay has not oxidised, there is a lack of vegetation on the newly exposed surfaces, there is a notice giving a warning that it is dangerous to climb the cliff, the promenade has been constructed as a sea defence, slabs of sandstone have fallen to the cliff base, at the east end of the promenade the natural rocks have been cemented to limit erosion; slumping, gullying and wind etching can all be observed in the cliff face.

10   Route

Follow the path to the East Cliff Pavilion (229365).

Activity: advantage should be taken of the view to summarise the main features of the Folkestone area, as seen so far, and to practise field sketching. With young children it may be best for them to complete and annotate outline sketches which have been duplicated for the purpose. Sketches should show the three parts of the harbour, the main part of the town, the railway line, the area of deposition to the west of the harbour and the bay to the east. (See Figure 53.)

FIG. 53   VIEW LOOKING WEST TO FOLKESTONE FROM CLIFF TOP PAVILION (G.R. 229365)

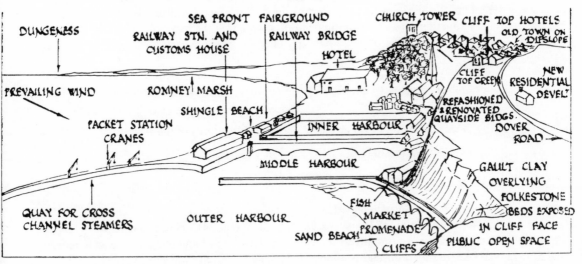

171

Comment: This is a convenient place for a late lunch, refreshments and toilet facilities are available in the Pavilion, the time should be about 1.15 p.m. If an earlier lunch is required a suitable point would be on the shingle beach west of the harbour. Here, the nearest toilet facilities are in the coach park and by the railway viaduct.

11   Route

Follow the path along the cliff top to Copt Point.

What to note: there are a number of old gun emplacements and the French coast can usually be seen. Where the coastline crosses the Gault Clay, a bay has formed, although it is not as large as it might be because Copt Point shelters this part of the coast from the prevailing wind and full erosive power of the sea. There is much evidence of slumping and the lack of vegetation in some parts shows the speed of erosion. All these facts emphasise the difference between the Gault Clay and the more resistant Folkestone Sands

12   Route

Follow the path down the Gault Clay foreshore.

What to note: at low tide Copt Rocks are revealed, these appear to be a wave-cut platform. If exposed, this feature could be used to emphasise the erosive power of the sea. There is some evidence to suggest that a backwash current occurs in the bay. In the distance the cranes and quays of Dover Harbour are visible just beyond Shakespeare Cliff. The effects of landslipping can be seen in the Warren.

The greatest dislocation and movement occurred in the middle layers of the Gault Clay and not, as is commonly believed, at the junction of the clay with the overlying beds of Upper Greensand and chalk. In the past landslips have been caused as a result of a combination of the following: i. groundwater making the Gault Clay moist, pliable, and slippery. ii. Prolonged periods of heavy rainfall increasing the weight of water in the overlying porous rock thereby compressing the softer plastic clay beneath. iii. Reduction of weight and pressure of seawater on the exposed surface of the Gault Clay during a period of very low Neap tides. iv. Undercutting of the cliff by the sea.

Activity:

(a) Collect samples of the three types of rock – Folkestone Sand, Gault Clay and Chalk.

(b) Draw and annotate a sketch of the view looking north to the Warren.

(c) Explain how a variety of rocks all appear at beach-level.

13   Route

A short but steep path now has to be followed. It starts from the point where the sea-wall begins (234373) and reaches the cliff top at 241366 near the Martello Tower. An alternative is to return to the cliff top above Copt Point and cross the golf course. Proceed to the footbridge overlooking the railway and the Warren (231374).

What to note: this is a good viewpoint for summarising. The situation of Folkestone on the Greensand dip-slope and the scarp of the North Downs can be seen to the south-west. To the north-east there is a further view of the chalk cliffs and the landslip débris in the Warren. A near-by notice states that the Engineers' Department of British Railways is responsible for maintaining the coastal defences in the Warren. Where the chalk cliffs rest on Gault Clay landslips have occurred.

A sea-wall and extensive concrete aprons have been constructed to prevent undercutting of the cliff. These defences reduce the likelihood of further movement; they are important because, at this point, the Dover-Folkestone railway runs perilously close to the coast. It crosses previously slumped material and has, in the past, been closed for long periods as a result of landslips. The most serious disruption of services occurred after the landslip of December 1915. It is probable that the upper part of the cliff is relatively safe as the chalk has no tendency to slip seawards because it dips towards the north.

Activity:

(a) An attempt may be made to sketch the view looking eastwards. Children will find this difficult owing to the considerable foreshortening of the view.

(b) The view westwards may be used for a few quick oral questions on the site and position of Folkestone.

Comment: The bridge could be used as a final summing-up point or to avoid a further climb the party could proceed westwards along the road to meet the coach at 235374 on the A 20. The coach could then take the party to the final viewpoint.

14 Route

Follow the path to 241377, a piece of ground just south of the 'Valiant Sailor'. Here there is an extensive view over Folkestone and across the Channel to France.

What to note: on the climb to the viewpoint you see a slight landslip in the Coombe. It has revealed solution pipes in the chalk. In summing up, stress can be laid on those features of coastal deposition and erosion which can be seen from this point. The main features of the position, site and function of Folkestone should also be considered.

Activity: the final task will depend on pupils' attainment. It would be useful to attempt a sketch map; younger children could be issued with an outline showing only the coastline, the route walked, the railway and the edge of the chalk. They could annotate this according to layout and function, marking in the following: harbour, fish market, extent of town, chalk ridge, Greensand Ridge and areas where the sea is adding to or taking away from the land. Photographs could also be taken for subsequent annotation.

Comment: For those dependent on public transport, buses run down to the town at ten-minute intervals. Coaches can pick up parties opposite the 'Valiant Sailor . If time is available it would be worth while to drive towards Dover and stop at the nearest viewpoint overlooking the town to note the contrast between the restricted valley site of Dover and the dip-slope of Folkestone. Rail access to Dover Harbour is less of a problem than at Folkestone and the harbour is much larger. These facts can be seen clearly from the walls of Dover Castle; this viewpoint could be used if the coach makes the return journey via Dover and Canterbury.

Alternative: the walk from Folkestone Harbour to the 'Valiant Sailor' could be done in reverse, thereby avoiding the ascent from Copt Point; the disadvantage is that the whole area is seen at the very start of the exercise and there is no good viewpoint for summarising at the harbour.

Follow up: sketches, maps, diagrams, photographs and written descriptions of the fieldwork should be produced. The postcards and photographs can be mounted and suitably annotated. It is best to undertake the write-up in two parts:
(a) the journey to Folkestone. In this case the pupils suggest key words to be used so as to ensure they concentrate their remarks on the important geographical features that have been observed. They should also be encouraged to draw cross-sections of Holmsdale, and east to west sections showing gaps in the Downs.

(b) the write-up of the fieldwork completed at Folkestone should include sketch maps of the coast showing the relationship of the direction of the predominant currents with the coast. A three dimensional model of the coastal area can be easily constructed using polystyrene tiles. This is of great help in further explaining the processes of marine deposition and erosion.

22

The Study of a Mountain Area

This chapter is concerned with the study of a cwm, a U-shaped valley and a prominent drift-covered Roche Moutonnée. In order to carry out fieldwork in a mountain area it is important that the following safety precautions be strictly observed:

(a) Always have with you a windproof anorak, spare warm clothing (especially gloves), and wear strong boots.

(b) Always carry emergency rations (and don't eat them en route!).

(c) Carry a compass, whistle, torch and small First Aid Kit — and leave word of your route. Also carry a short length of rope to help the nervous in difficult situations.

(d) Know where the local mountain rescue posts and nearest telephones are situated.

Planning the Route

(a) Estimate the time that it will take and make sure that you have sufficient hours of daylight.

(b) Remember that the weather can change very quickly and do not over-estimate your own stamina or ability.

(c) Treat the hills with great respect, particularly in snow conditions.

General Rules on the Mountain

(a) Don't take a short cut — invariably the path takes the safest, easiest and quickest way.

(b) Always descend the longer, more gradual side of a mountain; scrambling down steep rocky ground can be very dangerous.

(c) Never run, slide or flissade down a slope unless you can see a clear way to the bottom.

(d) Do not follow streams downhill. They may end in a waterfall.

The Study of a Corrie and Corrie Tarn. Cwm Idwal

Equipment required

Prismatic or 'Silva' type of compass. 2½" O.S. sheet SH. 65 1:25,000. 1" O.S. sheet 107 1:63,360. Foolscap drawing paper, sheet of hardboard, HB pencils, camera, binoculars.

175

Preparation which will help in understanding these features

(a) Extent of ice in North Wales during the last Ice Age.

(b) Introduction to the method of corrie formation.

(c) Origins of terminal and lateral moraines.

1 Route

From any field centre or camp site in Snowdonia the only way to approach the Nant Ffrancon valley is along the A5. There is a large lay-by along the shores of Llyn Ogwen at 659604 and cars or coaches can be parked here. Vehicles can also be left at the Ogwen Falls car park. Walk westwards along the main road until 654603, and at this point leave the main road and walk due south over the ill-assorted slabs of Ordovecian Volcanic rocks and join the footpath leading to the lower end of Llyn Idwal at 652607. This particular reference (at the gateway to the Idwal reserve) is a good point from which to photograph and sketch the backwall of Cwm Idwal.

What to note: whilst most of the excavation of the cwms took place in the principal glacial periods it is thought that small glaciers reoccupied them during the cold phase of the Late Glacial, ending 10,000 years ago. Many of the moraines found on the sides and lip of the cwm are due to this phase. They can be easily identified in the field. The majority of the cwms and their moraines face between north and east, for in such position there would be maximum shade and best opportunities for collection of snow drifted by prevalent south-westerly winds.

Activity:

(a) Photograph the headwall of the cwm and note synclinal folding of the rocks.

(b) Make a diagrammatic sketch of the cwm feature and annotate i. 'Devil's Kitchen', ii. Scree slopes, iii. Llyn Idwal, iv. Idwal slabs, v. Y-Grebin, vi. Height of the cwm headwall – measure this with a clinometer.

2 Route

Walk from 648598 to the foot of the headwall.

What to note: the Late Glacial glaciers descended at the lowest to 900 feet and in Cwm Idwal the corrie glacier almost certainly filled the Idwal cwm from 1,500 feet to 900 feet.

Activity: answer the following questionnaire:

(a) What are the approximate depth of the glacier filling Cwm Idwal?

(b) What was the direction of movement of this glacier?

(c) Besides the Névé ice, of what else was the glacier composed?

(d) What process of weathering during the late glacial period caused the recession and steepening of the headwall of this cwm?

176

(e) What is the name given to the erosive action which caused the deepening of the floor of the pre-glacial hollow?

(f) Draw a diagram at this point to illustrate the effect of headward sapping and rotational slip on the shape of the cwm.

(g) Why is it that post-glacial denudation has not removed the effects of the Ice Age?

(h) What type of weathering in particular could still cause erosion of these hard volcanic rocks at this altitude?

(i) What differences can you notice between the rock at the Headwall of the cwm and the material on the floor of the cwm and around the shores of Llyn Idwal?

(j) What feature at the head of Nant Ffrancon helped to contain ice within Cwm Idwal?

(k) What is the difference between the rock at the headwall of the cwm and the ill-assorted material on the floor of the cwm and around the shores of Llyn Idwal?

(l) What is the name given to the material on the floor of the cwm?

(m) What is the origin of this material?

(n) In what areas is this material most widespread?

Examination of Llyn Idwal

Llyn Idwal is sixty feet deep. Glacial excavation is the only effective method of cutting out such enclosed hollows. The ice of the corrie glacier was heavily changed with masses of rock derived from the gradually receding wall of the cwm, and pressed downwards by the weight due to its own thickness, and forward by the natural flow of the ice from the centre of accumulation above. This stone-laden cwm glacier scooped out a basin-like hollow because the erosive power would tend to be greatest where the ice was thickest.

As the ice melted and deposited a moraine across the Cwm, some material may have accumulated in the hollow that had been excavated in the rock partly filling it, or it may have been deposited on or beyond the lip of the rock basin, deepening the cavity in which water could accumulate when the ice finally disappeared.

Activity:

(a) Briefly account for the accumulation of water in this hollow.

(b) The lake is deepest near the headwall of the corrie. Briefly explain why.

(c) What factor has resulted in the apparent deepening of the glacial lake?

(d) What is the name given to the material which has dammed the lower end of Idwal Tarn?

(e) What is the origin of this morainic material?

(f) Make a sketch of the features at the southern end of the tarn and indicate the outlet.

(g) Llyn Idwal is very gradually getting shallower and smaller. Can you explain why?

(h) What evidence is there of settling around the shores of Llyn Idwal?

A Study of the Nant Ffrancon Valley

All the major Welsh valleys were glaciated during the Ice Age. Enormous quantities of ice moved down the former young river valleys, and removed inter-locking spurs as the ice ground on its inexorable course. The V-shaped pre-glacial cross-section became U-shaped, and what is often no more than a brook now meanders across the floor of an enormous glaciated trough. No Welsh valley better exhibits these features than Nant Ffrancon or Dyffryn Ogwen, which carries the A5 London—Holyhead road south-eastwards from Bethesda in Carnarvonshire.

FIG. 54   THE NANT FFRANCON VALLEY FROM Y-GLYDER FAWR

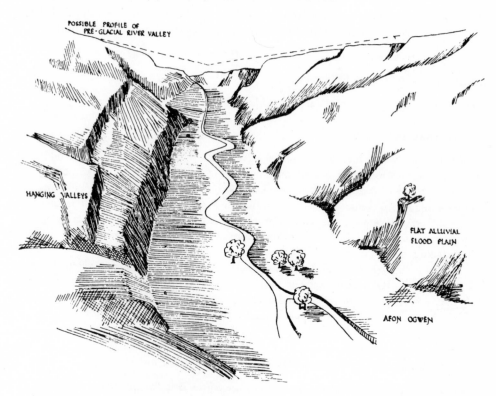

1   Route

Walk from Llyn Idwal to the Mountain Rescue Post at 650603 and from this point walk along the trackway until 645603 — this is a reasonably good vantage-point from which to look northwards down the Nant Ffrancon Valley. If the party is a Fifth or Sixth Form one, a much finer panorama of the Nant Ffrancon Valley can be obtained by walking to the summit of Y-Glyder Fawr. Although this would involve a few hours' walking the resulting view (assuming the visibility to be good) will be worthwhile.

178

What to note about the features of the Nant Ffrancon Valley: the Afon Ogwen now meanders across the flat alluvium-filled U-shaped valley. During the Ice Age stone-laden glaciers wore away the interlocking spurs and acted like gigantic rasps in clearing away the soil and weathered rock. This valley has a clear U-shaped profile and is particularly straight. This is because most of the valley traverses Ordorician sediments and the interlocking spurs of the pre-glacial valley have been almost completely removed.

The tributary valleys on the south-western side of Nant Ffrancon, for example, the stream at 632602 have gradients that are relatively gentle above 2,300 feet. Then below this they are quite steep so that the streams flow rapidly and give rise to picturesque but small cascades. At the sides of the U-shaped section of the main valley the streams plunge over the edge (see sketch) or flow in gorge-like notches as they make their way down the steep slopes that lead to the level floor of the valley in which the Ogwen flows.

Activity:

(a) Compare the base maps with the ground.

(b) Sketch the valley looking from 645,603 towards Bethesda. Although this is a relatively straightforward sketch duplicated sketches would help the less able.

(c) Annotate the sketch figure as shown.

(d) Compare the annotated sketch, with the sketch base map and the geology map.

Questions: i. Why are the sides of the valley exceedingly steep? ii. What evidence is there along the valley of small landslides? iii. The head of Nant Ffrancon is a steep 'dead-end' or cul-de-sac over which the Ogwen tumbles. Give one reason (related to structure) why this cul-de-sac feature is so steep. iv. Of what material is the flat floor of the valley formed?

2   Route

Walk of the back road northwards from 645603 along the track across rough pasture till one is immediately below the Rhaeadr Ogwen Falls at Pen-y-Benglog.

What to note: the waters of Llyn Ogwen drain westwards through a gorge cut in the cliff-like end of Nant Ffrancon to which it descends by the well-known Rhaeadr Ogwen. The gorge, really a gap in the original watershed, is thought to have been caused by water flowing beneath a glacier.

Activity: find answers to the following questions — using the map.

(a) What are the names of the mountains on either side of Llyn Ogwen?

(b) How high approximately is the waterfall at this point?

(c) What feature of the relief suggest that the pre-glacial watershed was at Rhaeadr Ogwen?

(d) Sketch the waterfall.

3 Route

Walk north-north-west from the position below the waterfalls for half a mile to 643610. From this point looking towards Y-Garn 6359 one can get a particularly good picture of hanging valleys with streams that tumble over the edge of the U-shaped valley side into the Afon Ogwen below the head of Nant Ffrancon.

Questions to be answered: i. why do these tributary streams have so steep a gradient? ii. Why are they called hanging valleys? iii. What evidence would we look for along the side of the valley wall to indicate the depth of the glacier that filled the valley?

Activity after returning from Nant Ffrancon: draw a cross-section from the head of Afon Perfedd 626618 north-eastward across Nant Ffrancon to the interfluve of the Afon Llafar at 646633. Mark in and label i. the approximate height of the main valley glacier, ii. the approximate position of the pre-glacial river valley.

The Study of a Drift-covered Roche Moutonnée

Among many other interesting examples of glaciation in North Wales is the study of ice-smoothed, scratched and plucked rock which because of its resemblance to the 'moutonnées' or sheepskin wigs of the eighteenth century has been given the name *roches moutonnées*. A particularly good example of this lies at the lower end of Cwm Dyli below Llyn Llydaw (Figure 55).

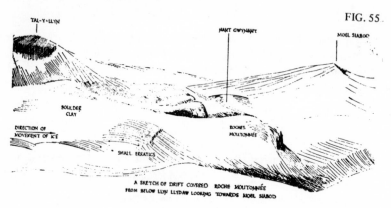

FIG. 55

A SKETCH OF DRIFT COVERED ROCHE MOUTONNÉE FROM BELOW LLYN LLYDAW LOOKING TOWARDS MOEL SIABOD

4 Route

Any vehicles could be parked at Gwastadannas 658536 and from there the party could walk to the footbridge across the Afan Glaslyn at 654537. This is a good point at which to stop and observe the downstream side of this particular roche moutonnée which is a quarter of a mile ahead at approximately 649537.

What to note: as the ice moved down from Cwm Glaslyn and rode over the hillocks and knobs of rock that lay in its way, it smoothed the rocky faces upstream, but on the downstream side it tended to pull away blocks that had previously been loosened by weathering along joints. Consequently a roche moutonnée is usually smooth on the side which faces up the valley but tends to be rough and craggy on the side facing

180

downstream. This is why a glaciated valley often looks strikingly different according to whether one is walking upstream or downstream.

Activity:

(a) Make a sketch illustrating the jointing of the rocks and the extent to which they have been plucked — roughly indicate the outline of Cwm Glaslyn and Llyn Llydaw.

(b) Why is it that the downstream sides of these rocks are apparently so well jointed?

5   Route

From the point below the roche moutonnée walk upwards to approximately 642536 and look east-north-east towards Moel Siabod. The view should be one similar to the sketch shown in Figure 55.

Activity:

(a) Sketch the landscape looking towards Moel Siabod as shown.

(b) Take note of the boulder clay drift and the many small erratics.

(c) Look for and note the striations on the bare rock above the Moutonnée — on any pavement feature the striae trend parallel to the movement of the ice.

Answer the following questions: i. What was the direction of the movement of the ice? ii. Why is the surface smooth on the upstream side? iii. What feature of glacial deposition gives the mountonnée an even smoother outline than it would otherwise have? iv. Under what conditions was the boulder clay deposited by the corrie glacier? v. What is the name given to the many small pieces of rock which are strewn in disorderly fashion over the surface of the boulder clay?

181

23

Fieldwork from Kindrogan
Field Centre:

(A) A Study of a Glaciated Valley

The aim of this chapter is to study landforms produced by a valley glacier and its meltwater; emphasis will therefore be on depositional rather than erosional features. It describes a method of fieldwork which involves accurate observation and recording in the field, and is designed so that the pupils look at all the evidence given by landform shapes, exposures of material and relationships between areas of land of differing shapes before attempting any analysis of their origin. However, it may be adapted so that discussion on the origin of landforms and their relation to one another takes place in the field.

Suitability

As the study involves only observation and description in the field, it is suitable for all pupils in secondary schools, although it may be necessary to do some simple theory on glaciation during follow-up. N.B. If weather conditions are bad it is not advisable to take younger age groups.
Time 9.30 a.m. 4.30 p.m.
Distance 6 miles.

Preparation by Teacher

(a) Seek permission to cross all forestry and agricultural land. *This must be obtained through Kindrogan Field Centre* (Warden: Mr B. Brookes), Enochdhu, near Blairgowie, Perthshire. Further information regarding courses and facilities at Kindrogan may be obtained either from the Warden or the Scottish Field Studies Association Headquarters, 141 Bath Street, Glasgow, C2.

(b) Walk round the route beforehand if possible.

(c) Become familiar with features of glacial and fluvioglacial erosion and deposition.

(d) Prepare duplicated base maps of the route, showing the main road (A924), Kindrogan Drive, Brerachan Water, Allt Fearnach, River Ardle and some prominent buildings. The 1,000 foot contour provides a useful guide line. These are to be used by the pupils at the final viewpoint.

Pupils' equipment: it is essential to wear warm clothing including gloves, hat and scarf during winter, spring and late autumn. An extra sweater should also be carried as it can be very cold in some exposed places. Strong walking shoes or boots must be worn and waterproof clothing carried at all times of the year. Each pupil must take a packed meal and drink.

182

FIG. 56

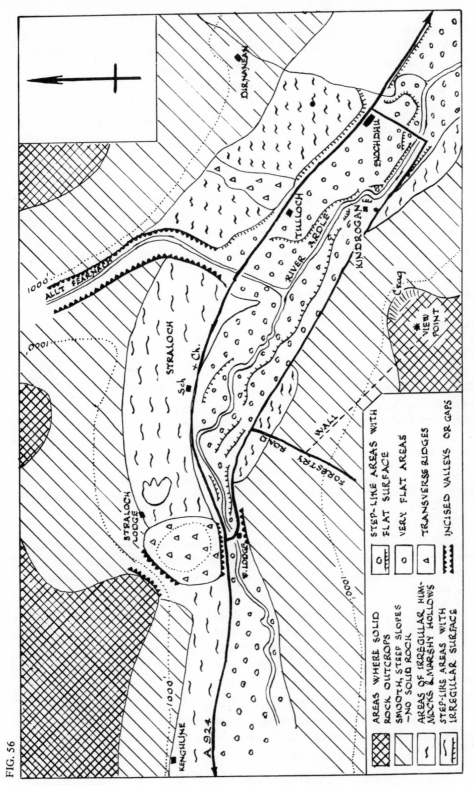

RIVER ARDLE GLACIATED VALLEY

Legend:

AREAS WHERE SOLID ROCK OUTCROPS	
SMOOTH, STEEP SLOPES — NO SOLID ROCK	
AREAS OF IRREGULAR HUMMOCKS & MARSHY HOLLOWS	∧
STEP-LIKE AREAS WITH IRREGULAR SURFACE	∫
STEP-LIKE AREAS WITH FLAT SURFACE	○
VERY FLAT AREAS	○
TRANSVERSE RIDGES	△
INCISED VALLEYS OR GAPS	⌄⌄⌄

Map labels: DIRNANEAN, ENOCHDHU, KINDROGAN, RIVER ARDLE, TULLOCH, VIEW POINT, Craig, ALLT FEARNACH, 1000', STRALOCH, Sch, + Ch., FORESTRY ROAD, WALL, STRALOCH LODGE, W. LODGE, KENGHLINE, A 924

183

Other equipment: 1″ O.S. Sheet 49 1:63360. Sheet of hardboard and clips. Pencils, rubber, field notebook. Geological hammer (one can be shared by 6 pupils). Polythene bags for collecting rock specimens and pebbles.

1   Route

The study begins on the front lawn of Kindrogan Field Centre (055629), which can be reached by public bus service from Blairgowrie or by private transport from either Blairgowrie or Pitlochry.

Activity:

(a) Observe and describe the shape of the land on both sides of the River Ardle. Note that there are very flat areas of variable *width* and *height* separated from each other by short, steep edges in a stepped arrangement.

(b) Examine the material in an exposure on the slope at the edge of the lawn, and note the size and roundness of the pebbles in it.

(c) Draw a field sketch of the area on both sides of the river, labelling the flat areas and steep edges and marking the exposure of material with details of shape and size of pebbles.

2   Route

Turn left out of Kindrogan Drive and follow the road to Kindrogan Bridge (063625).

Activity: compare the form of the valley bottom with that of the previous site. Note that it still retains flat-topped, steep-sided steps on both sides of the river.

3   Route

Continue to the main road (A924) at Enochdhu (063627) and turn left along it. After about one hundred yards enter the small quarry on the right hand side of the road.

Activity:

(a) Walk onto the top of the quarry. Take care to keep away from the cut edge of the quarry. Observe and describe the shape of the land into which it is cut. This is still step-like but no longer flat-topped as there are a number of hummocks and hollows on the surface.

(b) Determine the extent of land of this shape, and describe the shape of the land surrounding the quarry area, concentrating in particular on the steeper, smooth south-facing valley side in the Dirnanean area (065635). Assess the altitude of the break of slope between this surface and the surface into which the quarry is cut by reference to the 1″ map.

(c) Return to the quarry face and examine the material in it. Note that it consists of unconsolidated sands and rounded pebbles and boulders of numerous rock types. Collect samples for later identification. Remember to label where they came from.

184

(d) Draw a sketch of the section revealed in the quarry face noting particularly the sorting of the material into layers of different sizes. Some of the layers are horizontal and some dipping, indicating current bedding.

4 Route

Continue along the main road towards Straloch to the gate opposite Tulloch Cottage (056633). Go through the gate and up on to the ridge and follow the wall for about one hundred yards.

Activity:

(a) Describe the shape, height and steepness of the ridge and its transverse position in relation to the valley.

(b) Examine pebbles pulled out from any exposure and note their angularity. Collect a few specimens for later identification.

(c) Look up the Brerachan Valley and observe another larger transverse ridge immediately behind Straloch Lodge (036642) and the steep, smooth slope of the valley side to the north of it, which culminates in two rough, rocky outcrops. Assess the height of the outcrop of solid rock from the 1″ map.

(d) Draw a field sketch looking up valley showing the transverse ridge behind Straloch, the hill (Creag-an-tSithein) with rocky outcrops and the Fearnach valley entering the main valley from the north.

5 Route

Return to the A924 and walk up valley to the bridge across the Allt Fearnach (051639).

Activity:

(a) Note and describe the hummocky area to the right of the road ahead. Note the chaotic arrangement of the hummocks and their lack of uniformity of height, shape or orientation.

(b) Look up the valley of the Allt Fearnach and note the incision of the river, its swiftness, shallowness, width and the large boulders in its bed.

(c) Observe the flat land to the left of the road and the step profile of the field on the opposite side of the Brerachan Water.

6 Route

Follow the main road past Straloch Church to a point just past the school (044639).

Activity:

(a) Notice that the hummocks are still present to the right of the road and that they are separated by marshy hollows.

(b) Look at the exposure of material in a hummock which has been partly excavated just inside the field. Note the rounded pebbles and boulders of many rock types. Collect samples for later identification.

(c) Note the transverse ridge about a quarter of a mile ahead and the rocky outcrops of Carn Mor and Creag na Cuinneige on Creag-an-t Sithein.

(d) Draw a field sketch showing the hummocks, marshy hollows, and transverse ridge behind Straloch. Mark on the exposure in one of the hummocks and note the characteristics of the material on the sketch.

7 Route

Follow the main road to the point where the drive from Kindrogan joins it at West Lodge (036635). Go through the two gates opposite the drive, which are at right angles to each other. Walk over the transverse ridge and into the channel cut between it and Creag na Cuinneige. Enter the wood and follow the track to the point where there is a very large boulder on the floor of the channel (034640).

Activity:

(a) Using the 1″ O.S. map, note the relationship of the transverse ridge to the main valley and the position of the channel running sub-parallel to the main valley.

(b) Look at the characteristics of the channel and draw a sketch cross-section showing that the west side is cut in solid rock while the east is cut in unconsolidated material. Draw also a sketch long profile showing that the highest point of the channel floor is in the middle.

(c) Look at the large erratic boulder noting especially the water fluting on the downhill side.

(d) Collect samples of rock from the boulder and from the west side of the channel for later identification. Remember to label them.

8 Route

Leave the wood at the south end of the channel and climb up its west side. Walk parallel to the Brerachan Valley to a point (029641) from which Kenghline Farm can be seen.

Activity:

(a) Note the irregular hummocks and marshy hollows below, looking particularly for the long, narrow ridge trending towards Kenghline.

(b) Assess the altitude of the change from the hummocky area to a steeper, rock-strewn slope above, and the altitude at which solid rock outcrops, using the 1″ map for reference.

(c) Draw a field sketch or sketch profile showing the area between the Brerachan Water and Creag na Cuinneige.

186

9 Route

Cut back across the channel entrance and ridge towards West Lodge (036635), pausing on the top of the ridge to look at the Brerachan Valley.

Note: i. The flat floor of the valley upstream but narrow constriction downstream. ii. The characteristics of the Brerachan Water which is very sluggish, narrow, deep and meanders widely across its valley. Notice the filled-in abandoned meanders which have rushes growing in them.

10 Route

From West Lodge follow the Kindrogan Drive to the gates forestry road at 044635. Turn up this to the point where the wall crosses it.

Activity:

(a) Note the smoothness of the slopes and absence of solid rock.

(b) Look at and describe the material in the roadside exposure noting the angularity of pebbles, lack of sorting, and great variety of rock types. Collect samples for identification.

11 Route

Follow the wall to the rocky area just above Kindrogan Rock (048626).

Activity:

(a) Observe and describe the shape into which the solid rock has been worn. Note that parts are rough and parts are smoothed. Collect samples for later identification.

(b) Trace in detail the route followed during the day, almost all of which can be seen from this point. Stress the landmarks (such as buildings) which were seen en route and summarise the characteristics of all the areas examined during the day, making sure the pupils can see them all.

(c) Give out printed base maps showing the A924, Kindrogan Drive, streams, and settlements and the thousand foot contour. Make sure the pupils can orientate the map and can trace their route on it and on the ground as seen from the viewpoint.

(d) Allow the pupils ½–1 hour depending on their previous experience and aptitude, to map the distribution of the major types of landform which they have seen during the day (valley bottom steps, hummocky areas, transverse ridges, smooth, steep valley sides, rock outcrops). This can be done by use of symbols (see Figure 56, p. 183), colour or annotation. Check frequently that each pupil is mapping accurately by comparing land marks on the map with those he can see on the ground. It is essential that the pupil recalls what he saw at each stopping place as relief features are flattened when viewed from a height. Frequent reference must be made to the 1″ O.S. map.
N.B. If the cloud level is low it may not be possible to complete the map from the rock-top viewpoint, but it could be done back in the class-room by referring to field notes and the 1″ O.S. map.

Return to Kindrogan Field Centre via the path along the top of the rock (take care not to go too near the edge), and down the side of the larch plantation.

Follow-up:

(a) Neatening and filling in of detail on field sketches.

(b) Neatening and completion of the maps. A completed map could be drawn on the blackboard by a pupil who has done a particularly accurate one.

(c) Identification of samples of solid rock and pebbles.

The solid rock is foliated schists containing either hornblende (dark green-black) or mica (platy and shiny) and garnets (red-brown crystals). The pebbles are of much greater variety including schists and also pink and white granites, porphyries, and quartzited as well as other less common specimens. A comparison with a ¼″ Geological Map will show that granites, for example, must have come from at least seven miles away.

(d) An analysis of the formation of the various types of landform and their relationship to one another is now possible using evidence from completed maps, field sketches, field notes on shapes of land, exposure material etc.

A good starting point for this analysis is the two transverse ridges. Their position across valley and the unconsolidated angular pebbles of very many types suggest that they are terminal moraines.

Other evidence of glacial deposition is found on the smooth, steep upper valley slopes where bedrock has been masked by a cover of glacial till which contains unsorted material with angular fragments. Till is a local Scottish word which merely means stony ground. It has now replaced the older and more ambiguous term 'boulder-clay' since material formerly so described need contain neither clay nor boulders.

From this one can lead on to a discussion about the origin of the material which forms these deposits. Some of the rocks must have been transported over quite long distances by ice while others are of more local origin. Evidence of glacial erosion in the form of ice plucking and ice-smoothing can be seen on the surface of Kindrogan Rock. A comparison of the Brerachan and Fearnach valleys shows that the former was occupied by the dominent glacier which overdeepened it more than the latter, which thus forms a slightly hanging valley.

The two terminal moraines are breached where the present streams cross them, and that at Stralock also where the big channel runs across its northern end. The evidence of water-fluting on the downhill side of the erratic in the latter shows that the channel was eroded by water, and that in this case it must have flowed uphill. This could only happen because the meltwater was flowing under hydrostatic pressure under ice.

As well as acting as an erosive agent, meltwater was responsible for deposition, sometimes under ice. As a result of this the valley has many kames (irregular hummocks of rounded material formed either along the margins of melting ice or in crevasses in the stagnant ice as water flowed along or through it); eskers (long, narrow ridges formed by deposition in the bed of streams flowing in tunnels in or under the ice); kettleholes (hollows left when isolated blocks of stagnant ice, which had been coated by meltwater deposits, finally melted); and outwash fans (current-bedded deposits formed where meltwater poured through breaches in a moraine). All these contain material originally excavated by ice but which became rounded while being transported in meltwater.

The valley bottom is infilled to an unknown depth and the river is now re-excavating these deposits to form meander terraces, which are flat-topped and steep edged and consist of unconsolidated rounded deposits.

(B) A Study of a Highland Estate

Aim

To understand agricultural economy in relation to a marginal environment.

Suitability

Any secondary age group.

Preparation

Teacher:

(a) Contact an estate owner.

(b) Adequate preliminary reconnaissance is necessary.

(c) Prepare base maps and questionnaires as required. Scales of one inch to the mile to show the estate boundary and physical features of the site (see Figure 57 below), and of six inches to the mile for mapping features of the lower areas, are particularly suitable.

Pupils: locate the estate on a 1″ O.S. map and carry out preliminary map reading with reference to its position and physical site.

Sample study

Dirnanean, Enochdhu, Blairgowrie, Perthshire.
Mr F. K. Balfour.
This study concerns an estate of 5,000 acres situated in a glen in the Southern Grampians. It is fairly typical of estates of this size in the area and the principles underlying its economy can apply equally to similar farms.

Teachers wishing to use this particular example *must* arrange their visits through the Warden, Kindrogan Field Centre, Enochdhu, Blairgowrie, Perthshire.

Grid references refer to the 1″ O.S. Sheet 49.

Route

The fieldwork can be done by following the Glenshee track from Enochdhu (0662), branching off into selected fields, depending on the distribution of crops and animals.

189

FIG. 57

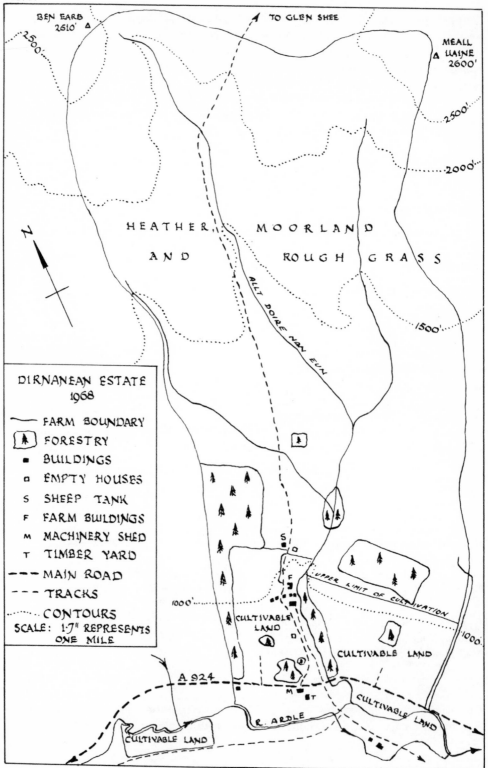

BEN EARB
2610'

TO GLEN SHEE

MEALL
UAINE
2600'

2500'

2500'

2000'

H E A T H E R   M O O R L A N D

A N D   R O U G H   G R A S S

N

ALLT DOIRE NAN EUN

1500'

DIRNANEAN ESTATE
1968

—— FARM BOUNDARY
FORESTRY
■ BUILDINGS
□ EMPTY HOUSES
S SHEEP TANK
F FARM BUILDINGS
M MACHINERY SHED
T TIMBER YARD
– – MAIN ROAD
- - TRACKS
···· CONTOURS
SCALE: 1·7" REPRESENTS
ONE MILE

S

UPPER LIMIT OF CULTIVATION

1000'

F

CULTIVABLE
LAND

□

CULTIVABLE LAND

1000'

A 924

M
T

CULTIVABLE
LAND

R. ARDLE

CULTIVABLE
LAND

Remember to keep out of fields where there are very young animals in spring. Always walk around the edge of fields containing young animals in spring and close all gates.

It is not necessary to go beyond the point on the track from which the watershed, which forms the north-eastern boundary of the farm, can be seen and from where the character of the 'Hill' can be observed (approx. 073643).

It is suggested that observation and discussion of the field evidence of the farm economy takes place on the way up and activities such as mapping and sketching are carried out while returning to Enochdhu.

What to note

1   The Fields

The Field Pattern:

(a) In the valley bottom and on the lower slopes, the fields average about 10 acres. They are usually stone-dyked or post and wire-fenced and reflect late 18th or 19th century enclosure.

(b) The heather moorland is ring-fenced and otherwise unenclosed or divided relatively recently by post and wire fences into large units.

The Use of the Fields: Cultivatable Land

(a) *Good grassland:* This is usually temporary, being ploughed up and re-seeded at fairly regular intervals in order to maintain quality and is, therefore, an important cultivated crop.
i. Note the altitude of the boundary between the improved grass and the poorer permanent grass or heather moorland.
ii. The good grassland is used partly for grazing and partly for the production of fodder for winter.
Autumn – used for grazing store calves and lambs prior to their sale to lowland farms for fattening.
Winter – grazing of cows and their new-born calves. The cows are given supplementary food.
Spring – cows and young calves and sheep and new-born lambs graze.
Summer – three-quarters is used for hay and silage for winter fodder, and a quarter of the area for grazing.

(b) *Arable:* i. note the altitude to which arable crops are grown.
ii. Small acreages of oats and root crops are grown for winter feed.
iii. In the past, the growing of these crops has been much more important and there is much evidence of this in the buildings, e.g. the long drier where oat sheaves were hung over wires to dry; granaries in the farmstead. Until recently, the farms were much more self-sufficient in winter keep as a smaller number of cattle was kept and the number of animals over-wintered depended largely on how much fodder was grown. Now food, especially proteins and barley straw, is either produced on a lowland farm belonging to the same owner or, as in the case of Dirnanean, is bought in. The expansion of the breeding cow stock has led to the intensification of grass cultivation.

191

Forestry

(a) There is older forestry on dry, sandy, uncultivatable hummocks (kames). Some of this is semi-natural birchwood and some mature conifers, especially Scots Pine and Norway Spruce.

(b) Blocks of coniferous forest have been planted as a long-term investment. These provide shelter for animals and allow the economic employment of men who are then available for other farm jobs in time of great pressure, e.g. shearing, hay-making.

(c) Small areas of young conifers on otherwise unusable slopes, e.g. terrace edges, steep tributary valley sides.

Rough Grazing and Heather Moorland

(a) *Permanent grass:* this is found on some steep, unploughable slopes below 1,200 feet and is dominated by coarser, poorer grasses than those of the temporary leys.

(b) *The Hill:* this occupies about four-fifths on the farm. In places, especially the lower parts, rough grasses predominate, but at higher elevations they are gradually replaced by heather, which is carefully managed by burning to encourage the growth of young shoots. This is done in early spring when very small patches are fired so that the flames can be easily controlled and to allow a scattering of patches of new growth. In early spring look out for fires and at all times of the year note the patchy colouring of the 'Hill'. The light patches represent young heather while the darker areas are more woody heather shrubs.
i. The Hill is used for sheep and cattle grazing at all times of the year. As heather is an evergreen shrub, it provides food all year.
ii. The area is also of great sporting importance. The income from sporting activities i.e. grouse, deer, hares, etc. provides a useful adjunct to the economy of this estate as well as a substantial amount of meat. On some estates, revenue from letting out shooting right is considerable.

2 The Animals

The whole estate economy is dominated by extensive grazing geared to the production of store calves and lambs for sale.

(a) *Beef Cattle*
Highlanders (brown; long, shaggy coats; long horns).
Beef Shorthorns (white or red).
Aberdeen-Angus (small; black; hornless).
Luing (a stabilised cross between the Highland and
Beef Shorthorn, which is hardy and matures quickly).
Other crosses are quite common.
   A few stores are now over-wintered in cattle courts and are fed on home-grown silage, and hay with protein food bought in.

(b) *Sheep*
Scotch Blackface (small; horned; long, coarse wool).

192

The lambs are born in late April and May and, by September or October, are ready for sale at Perth. A substantial part of the income from sheep is provided by wool.

3 The Farm Buildings and Houses (steading)

(a) The owner's large house is set in a large ornamental garden and nearby is a walled garden used for the production of vegetables. Around are a number of cottages.

(b) The farmstead is built in a typical Scottish square plan around a central courtyard. Note that it is constructed of local stone. Newer machinery sheds and cattle courts for in-wintering animals are later additions. Near the farm buildings are a number of stone-built, fully-equipped and modern farm-workers' cottages.

(c) Near Enochdhu (062628), there is a large machinery shed and also an electrically-powered timber yard where fencing posts are cut, using home-grown timber.

(d) At the edge of the Hill is a sheep 'fank'. This consists of pens, a shearing clock, a wooden support for a wool sack and a sheep dip.

4 Labour and Mechanisation

As labour is increasingly difficult to obtain in these upland areas, the estate has become highly mechanised. Note the partially ruined cottage at Braegarrie (068643).

(a) There are a number of tractors and tractor-drawn items of machinery in the sheds around the farmstead or at Enochdhu or in use in the fields.

(b) The farm has a number of other vehicles, e.g. Land Rovers; cattle wagon.

(c) Note other mechanical aids, e.g. power saws.

5 Other Important Features of the Economy

Many other factors affecting the economy could be studied where information is available. These include the use of lime and fertilisers, marketing and, perhaps most important of all, deficiency payments. Information about current deficiency payments, can be obtained from the National Farmers' Union.

Activities:

(a) Drawing of field sketches and plans of farm buildings. Also mapping of land use.

(b) Investigation of the degree of dependence of land-use on physical features. Physical factors, which should be examined in greater detail and perhaps mapped, include: i. site − aspect, altitude, slope. ii. Landforms − meander terraces, hummocks of glacial or fluvio-glacial origin, steep valley sides, valley bench, mountains, etc. iii. Drainage − N.B. artificial this drainage of the lower areas is just as important as natural run-off. iv. Soils and their parent material, which is usually alluvial or glacial in origin. The soils are rarely derived from the underlying schist rocks. With more advanced pupils, analysis of soils can be undertaken in greater detail looking especially at texture, porosity and pH value. Note that the cultivated areas have more alkaline soils than those under permanent grass, largely due to liming.

24

Fieldwork and Geology

1 Studying a Quarry

Aims

(a) To demonstrate clearly the geology of an area and to study the geomorphological process which have affected it.

(b) To show the economic use of rocks.

(c) To discover evidence of geological phenomena and illustrate dependence of the present on the past.

Suitability

All students can gain much from a properly directed observation of quarries, etc., but remember the need for

Safety Precautions

(a) Always find out when the next blasting will occur. Blasting usually takes place at 12 noon.

(b) In working quarries keep away from men working and from machinery whether operating or stationary.

(c) Beware of falling rock fragments and refrain from hammering under overhanging bluffs.

(d) Do not climb quarry faces except by way of recognised pathways.

(e) Beware of injuring other people with flying chips of rock when hammering and of dislodging masses of rock on to people below. Avoid working immediately below anybody.

Equipment

A selection of the following: i. Geological hammer (forged steel only), trowel, entrenching tool, bolster (geological chisel); ii Soil auger; iii. clinometer; iv. compass; v. tape measure; vi. haversack, newspaper for hardy specimens, cotton wool and boxes for delicate material; vii. adhesive tape for marking specimens; viii. geological maps, 1″ and 6″; ix. O.S. maps, 1″ 2½″, 6″, 25″; x. suitable clothing (hard wearing, waterproof); xi. boards, clips, elastic bands, etc. for notes and sketches; xii. pencils, black and coloured, and rubber; xiii. hand lens; xiv. binoculars; xv. camera; xvi. First Aid kit.

Preparation

Teacher:

(a) Always get permission from land or quarry owner before visiting the site.

(b) Study maps, books and geological memoirs carefully to equip yourself with an excess of information.

(c) Make the excursion personally and check all features which are to be noted on the excursion, time taken, etc.

(d) Note bus and train times, places suitable for rest, lunch, toilet, etc. (Consider the use of other schools for some of these.)

(e) Prepare hand-out material or instruct class in making preparations, e.g., Base maps showing area, contours, spot heights, drainage, rock sequence, landmarks, etc.; questionnaires; route sheets; cross-sections.

(f) Brief pupils adequately about tasks, activities, etc., before the excursion and give an introductory lesson to fit the visit into perspective.

Activities:

(a) Know where you are: i. orientate the map with compass, landmarks, sun, etc., as soon as you arrive; ii. find grid reference.

(b) Observe attitude (configuration) of beds: amount and direction of dip (angle between surface and bedding plain), true dip (angle between horizontal and bedding plain), height of quarry face.

(c) Observe kind of rock: sedimentary, igneous, metamorphic, and what type of these main classes: bedded, massive or intrusive.

(d) Note texture of rock: course, fine, conglomeratic, brecciated, current, bedded or graded.

(e) Observe structures: folds and faults.

(f) Collect specimens and fossils and wrap carefully after labelling with adhesive tape to show grid reference and position in face as near as possible.

(g) Make cross-reference notes in notebook giving information about circumstances of finding the specimen. A field sketch and/or a photograph should be added.

(h) Do not throw specimens away carelessly because they may confuse future students and teachers.

(i) Find out the uses of the rocks by enquiry, observation and reference books.

(j) Note techniques of transportation and quarrying.

(k) Visit the processing plant and find out the methods used to make the raw material more useful to mankind.

Comment: Most of what has been suggested so far applies particularly to secondary and advanced students. The most rewarding work for the majority of school-children lies in the study of the raw material *in situ*. Processing plants are often complex and noisy. A simple explanation of what goes on may be as instructive as a tour of the plant.

2   The Geology of a Hole

Any excavation into the earth is of interest to geologist, miner, engineer, archaeologist, etc. It is surprising how many undertakings ultimately need to dig into the ground. The resulting holes, shafts, etc. range from the smallest roadworks to quarries, oil and water bores, mine shafts and largest, deepest of all, the MOHOLE which the Americans have drilled for the purposes of scientific investigation. Holes, then, both large and small are worthy of study because:

(a) They reveal rock and soil profiles in a precise location.

(b) Collation of sections from several holes over an area can give information about the geology of the area which may add to or clarify the geological map (e.g. precise identification of the edge of a river terrace).

(c) Some information may be gathered about water supply, drainage, water holding capacity of certain rocks, etc.

(d) In urban areas it is often difficult to discover geological details due to built-up areas, hence temporary roadworks or wholesale developments (e.g. Southgates Underpass Scheme, central Leicester) often provide an opportunity to find out more. Many schools may be situated near to such roadworks.

(e) Large, economic holes (quarries) constitute a positive threat to surrounding areas e.g. to nearby houses (the Potteries), to agricultural land, to landscape quality, amenity value etc. Hence they need to be carefully recorded and mapped at all stages.

(f) Holes are of particular value to the archaeologist, they may produce unexpected and important finds especially in city areas.

Method

(a) The *Leicester Mercury* lists current roadworks each weekend. Here are the main ones listed for November 1967: Southgate Underpass area – West Bridge. Near Millstone Lane (sewer workings). Lower Hill Street near Charles Street (sewer works). St. Matthews Redevelopment Scheme (off Humberstone Gate) etc. But ordinary observation will soon reveal that this is the high season of hole-making in the city.

(b) Try to visit several holes. It would be a good idea to make record cards for each one. These would show the following:
i. Exact location hole. ii. Purpose. iii. Age and duration. iv. Hole data:
Dimensions (approx.) – sketch plan from above?

Depth dimensions – section sketch noting which face you are looking at (i.e. North, South etc.). Note colours, textures, bedding, harder rock bands, soil horizons, evidence of minor faulting, ripple marks, etc.

196

State of bottom – depth of water (if any). Is pumping going on? Nearness to river, canal, etc. Are the sides battened up or collapsing?

This information could be arranged briefly (a code might be developed) on a card together with small sketches.

(c) Plot holes visited on a street map in relation to geology. (Follow-up work could be done on the interference caused to pedestrians and vehicular traffic flow by holes.)

A soil monolith could be made later as a model of the hole section studied. If this was to be done soil samples would have to be collected on the visit.

Completion of work

If time permits three concluding activities can be attempted:
(a) Compare findings to borehole sections given in Sheet Memoirs (in Reference Library, Leicestershire Room, Belvoir St.).

(b) Complete index data cards for each hole as basis of permanent collection.

(c) Write a descriptive comment on the geology of the small area covered from your own observations.

Appendix: 1

Assessing Fieldwork for the C.S.E. Examination

During the last few years fieldwork and project work have become more widely accepted and practised as methods which, rightly used, are of great educational value at all levels of geography teaching. They are valuable and worthwhile because they involve each individual pupil in observation, estimation, practical work, and use of reference material. All this involves the teacher in preparatory work, in addition to the formidable task of marking and assessment when pupils' assignments have been completed. Many teachers have had very limited experience of this kind of assessment for examination purposes, and may find it useful to adopt the following method of procedure which is suitable for application to the assessment of fieldwork and project work at all levels.

For examination purposes many teachers have to arrange the names of their candidates in merit order. Some have to suggest the grades which they consider should be awarded to candidates for projects and fieldwork. Prior to commencing this work the teacher has to find the answer to three questions:

(a) How many pieces of work have to be assessed?

(b) Have adequate arrangements been made to ensure that all candidates submit their work in good time so that the total school entry is available when assessment commences?

(c) How much time and how many marking sessions will have to be allocated to the task of assessing all the work?

A group of between thirty and sixty scripts is the ideal for assessment purposes. With a larger group of candidates it will be necessary for several teachers to assess the work of the candidates in their charge and for the head of department, acting in consultation with those teachers, to prepare the co-ordinated order of merit and grade assessment for all the candidates at the school. Assessment of small groups of less than ten candidates also presents problems, especially during the early years of the examination's existence while individual teachers are learning to associate regionally acceptable grades with the standards of achievement of their own candidates. Assessment problems are accentuated if the candidates produce work of a very similar standard and/or if the teacher's ideas about criteria and attitudes to marking are not in accord with those of the examiner. External assessors and moderators will no doubt take a particular interest in these groups.

Some advice about criteria and attitudes to marking is given in the H.M.S.O.

198

publication *C.S.E. Bulletin No. 3, An introduction to some techniques of examining*, but due to the novelty of such a large scale operation some teachers may still feel that they have not been given enough detailed advice. It is with their needs in mind that the following suggestions are made:

(a) Study the syllabus and any other advice on criteria issued by the examination board.

There is general agreement that assessments should be based on the giving of positive credit for good points. Of their very nature untidy work and poor spelling penalise the candidate without the assessors having to make any mark deductions; conversely an assessor of a geographical project or piece of fieldwork should not give additional credit because the work is well bound, artistic, or neatly written in perfect English. Candidates who have these accomplishments will more easily be able to convey the geographical concepts on which they are being assessed. Only through the higher standard of geographical understanding shown by the candidates should they receive credit. Work should be marked for quality rather than quantity; sheer volume of output is often a very poor guide to subject ability. Obviously undigested facts and unedited copying from guide books and text books are not good geography.

A subject teacher who has doubts about grade standards and criteria of assessment would be well advised to contact colleagues who teach the same subject in neighbouring schools. Some of these teachers will probably be more experienced and be able to pass on valuable information about the fieldwork potential of the local area. This kind of information will be of particular value to a teacher who has recently moved into the district. In some areas it might be possible to arrange for the subject teachers from several schools to take a few sample scripts to a local meeting so that corporate assessments could be made. As a further aid to the establishment of common standards examination boards may arrange displays of graded sample scripts. These displays will be of particular value to teachers who have only recently qualified.

(b) Once the project and/or fieldwork folders have been completed and handed in by all the candidates the teacher can commence the work of assessment. Where possible the initial marking session should continue until all the work submitted has been looked at. This gives a good overall picture of the complete range of the work submitted and by making this initial assessment in one session variations of marking standards are minimised. When estimating how much time to set aside for this initial assessment, about five minutes should be allowed per script plus about ten minutes for reassessment. The most convenient technique to apply is 'coarse grading'. Quickly scan the first candidate's work, avoiding detailed study of the whole but looking carefully at one or two items in the main body of the work. Formulate an assessment on the basis of 'above average', 'average', 'below average'. Go through this procedure with all the work so that you finish up with three piles of scripts corresponding to the three coarse grades. You should then take a second look at the scripts at the bottom of each pile to check that your standard of assessment has not altered during the course of the session. Where necessary reassess the work which you first looked at. The three piles of scripts need not have the same number of items in them; the middle pile will probably be the largest of the three.

(c) Go through the work in each of the three piles doing more sampling and subdivide each pile so that you finish up with six ability groups. If a mark scheme is available it is probably not intended that it should be used as a rigid guide as in GCE marking. Its main value at this stage would be to help the teacher to differentiate between the relative ability of the candidates within each of the six groups. It could well be that three or four candidates are deemed to have submitted work of equal merit, no attempt should be made to create artificial divisions between them; in an order of merit these candidates should all be entered on the mark sheet as being of equal calibre. It would probably be useful to have a table set out, as in the example given, to be filled in as the teacher goes through the work for the second time so that reference may be made to it in the drawing up of the order of merit. The density of ticks entered on the table will be a guide to ability.

	Little	Moderate	Much
'Guidebook' style writing			
Irrelevant material			
	Above average	Average	Below average
Original observations (Written)			
Original observations (Illustrated)			
Relevant illustrations (Photos, Drawings, Maps)			
	Good	Average	Poor
Explanation of observed facts			
Initiative and originality			
Organisation and presentation			

(d) In the final review before drawing up the order of merit the geographical content of the work can be tested by careful and detailed marking of a map or diagram and of several paragraphs selected at random in the main body of the work.

If the Board require a mark or grade this should be considered after the order of merit has been established. The six divisions made during the coarse grading process can with only slight adjustments be used for the examination grade assessments. If marks are required the full range should be used and the teacher should make it clear that this has been done. With only a few candidates this is not possible, but if there are more than thirty candidates the best should be awarded maximum marks or near maximum marks and the worst minimum or near minimum marks. Although familiar with the shortcomings of impression marking, especially if it is not applied in one continuous marking session, many experienced teachers allocate an impression mark out of twenty for pupils' attempts at sample 'O' level questions and other school work. If marks have to be used the choice of twenty as the range for such impression marking is probably the most suitable for assessing projects or fieldwork. It gives a reasonable number of mark positions over which to distribute the candidates yet there

200

are not so many positions as to make it difficult for the external assessor or moderator to make any necessary adjustments between school and regional standards by calling for samples and/or by using the rest of the examination as a reference unit. Where the teacher does not use the twenty marks for impression marking but tries to break them down and allocate five out of twenty for sketch maps, five out of twenty for observations, etc., two serious problems arise. In the first place there is a risk that few teachers will give nought out of five or five out of five on each section. As a result the total marks for the candidates will run from four out of twenty to sixteen out of twenty. This has the dangerous effect of reducing the mark range by eight out of twenty. Secondly there is a very real danger of regression or 'bunching' exerting an unreasonable influence. This can best be illustrated by reference to the chart example shown here.

Candidates	Maps	Observations	Reasoning	Layout	Total
Smith	3	4	2	3	12/20
Jones	2	4	4	2	12/20
Bloggs	4	3	4	1	12/20

In each case the candidates apparently produced work of varied standard on particular sections of the mark scheme but they all achieved the same total mark. This kind of pattern reproduced in many examination centres would result in large numbers of candidates of differing ability being awarded the same mark. Because of these statistical difficulties Boards will no doubt quickly adopt the teacher's order of merit and estimated grades as the chief criterion when assessing projects and fieldwork. When this becomes general practice teachers will readily appreciate the value of the 'coarse grading' technique. Meanwhile where coarse grading is used and marks are allocated subsequently, on the basis of the order of merit produced by coarse grading, the resulting assessment will be more consistently accurate than one derived from normal impression marking.

Appendix: 2

A Self-help Approach to Fieldwork Preparations A Case Study

In recent years much consideration has been given to the problem of communicating new methods and ideas to practising teachers. Lectures, day conferences, and periods of in-service training are useful but are too often only a means of transmitting information to a passive audience. Unless the teacher is actively participating in the formulation of new ideas and methods he may not feel committed enough to apply them to his own teaching. It is one thing to talk about teaching – it is another thing to teach a class of thirty or more children in an enlightened way.

The Nuffield projects demonstrate what can be done to inject new life into the teaching of science by a group of co-operating teachers. The Schools Council has established lines of communication between teachers wishing to work together on devising methods of curriculum development. The Council's Working Paper No. 10, *Curriculum Development: Teachers' Groups and Centres*, sets out the organisation for the creation of local groups and centres. This appendix describes the work of an association of geography teachers, who meet to assist themselves in developing ways and means of teaching local geography. The evolution of this self-help approach to fieldwork preparation might provide a useful guide to the development of groups elsewhere willing to work on this aspect of the school curriculum.

The group originated as a result of studying the requirements of the Certificate of Secondary Education geography syllabus. All well and good for the C.S.E. Bulletin No.1 to state that pupils at the end of a five year course should be 'observant of the natural and human landscape around them .... progressively precise in this observation and the recording of it, and increasingly seeking explanation of what they see'. It is quite another matter for teachers to put this into practice with their pupils while coping with the day-to-day demands of teaching. There are still many teachers who have not received a training in the use of fieldwork, and fieldwork is necessary if the pupil is to develop his powers of observation. Other teachers may have no particular flair for developing their teaching in this direction. Indeed, the paradox is that although fieldwork has been for some time the growth point of modern geography teaching, yet there are still many teachers who have not caught up with that

202

development. The situation is made more imperative because all available evidence suggests that fieldwork may become a compulsory part of the C.S.E. syllabus for geography: several Boards have already made it so.

The practice of fieldwork implies that the teacher must have a knowledge of the local area. In addition the C.S.E. syllabus rightly lays emphasis on the classroom study of the region in which the school is situated. Even teachers trained in fieldwork methods, but newly arrived to the area, find it hard going to acquire the necessary information concerning the local geography. The effort is made more difficult because reference material rarely exists which is suitable for making an approach to the new kind of teaching implicit in the better aspects of the syllabus. Textbook material is usually far too inadequate in coverage; other sources of a more academic kind, where they exist, need a great deal of quarrying before information is gathered that can be usefully presented to the C.S.E. pupil. Hence the teacher finds himself in a weak position to prepare creative lessons at a time when his most creative approach to the teaching of the subject is required of him.

It was because of these difficulties that a number of teachers, working under the auspices of the University of Leicester School of Education, met to discuss the teaching implications of the C.S.E. syllabuses being shaped by the panels affecting Leicester and the County. During these meetings constant requests were made, 'But how do we *do* fieldwork?'; 'How do we teach about Leicestershire when we haven't got the material to hand?'. Many ideas were put forward to remedy this. Some of the more experienced teachers offered to lead fieldwork excursions demonstrating the relevant techniques to the less experienced members of the group. Several expeditions were attempted but it soon became clear that although attendance at these meetings was quite large, the groupings of the people involved did not carry enough impetus to make further headway. Many teachers came to receive information but were not ready to work out excursions for themselves. Consequently the efforts made to disseminate fieldwork methods and knowledge of the local area remained on a semi-organised and desultory level.

The situation was like this until one member proposed a way of solving the problem of inaction which had overcome the meetings. The new approach outlined was that small groups of the teachers should work on a chosen area of Leicestershire either where their school was situated or in a place of intrinsic interest to the individuals concerned. The teachers would find out all possible information concerning the area of their choice either through studying written source material or by finding out for themselves through fieldwork. Such an area would form a small sub-region of the County. Thus the task of gathering teaching material for the whole County was broken down into manageable proportions for each group. The write-up of each area would contain class material and fieldwork exercises and this would be distributed amongst the members and to teachers unable to take part in the 'spadework' involved. At least this would provide the material for the teacher to base his local geography lessons on as well as giving a guide to the fieldwork excursions that could be organised in conjunction with descriptive information.

The initiator of the idea drafted an outline of the first contribution taking the Vale of Belvoir as the example. It was this 'model' which first polarised the interest and effort of a number of teachers who then voluntarily came together to work on particular areas of their choice. Some of these teachers were fresh from college; others heads of geography departments. The total active membership dropped to sixteen although some fifty teachers remained on the mailing list and received information concerning the progress of the work. The important point is that these sixteen teachers began actively to co-operate on a project: it was no longer the case of a group being led. The 'model' had provided a means for structuring the enquiries and the reporting back that the members could follow. The agreed goal of attempting to circulate the completed work gave an extra fillip to the activity. A measure of the members' enthusiasm is shown by the fact that the group, named the Leicestershire Association for Local Geographical Studies on its foundation in January 1966, meets at fortnightly intervals during term time.

Once having structured the kind of activity that the Association was to pursue, it became equally necessary to structure the group of teachers as a working body. It was considered of great importance that a hierarchical structure should not be allowed to develop. At first, even, the newly constituted Association worked without a permanently elected chairman, although one was elected when it became apparent that continuity of effort could not be made without one. Another important office, that of secretary, was created, his role being to record the decisions made by the Association. He does not keep a verbatim minute of the proceedings. The record of decisions has a self-regulatory effect on the members, acting as the collective memory for the obligations undertaken by them. The secretary also retains manuscripts of work as they are submitted. These are then readily available for reference by active and 'sleeping' members alike.

At each meeting a task is agreed upon which is to form the basis for an agenda at the next meeting. This task has to be completed before that meeting and usually consists of members presenting their draft manuscripts, or a written statement of the kind of investigations and material their contribution will contain. These are then presented at the meeting for discussion. Early on the important principle was agreed that each member must be prepared to give and take constructive criticism of the most stringent kind. This can be a most salutary experience. Thus no one, whatever their status or experience, is in a position to pontificate.

Another factor in maintaining co-operative inter-personal relationships has been the refusal of the Association as a whole to form a small editorial board. This, it is felt, might make a clique of two or three dominant members who would, through the nature of their office, dictate somewhat to the other members. It is the whole body of active members who make up the editorial board, provided they have read the manuscripts under discussion. Another benefit from this is that teachers who are unable to take an active part in the work, but who might occasionally attend, can voice their opinions on the work under review. This keeps the Association in touch with the needs of teachers outside the immediate active membership.

Of course this method of editing each other's work is a long-winded one, and clear-cut decisions can take a long time to be made. Indeed some members were at the outset opposed to this method for that reason. Undoubtedly it is unlikely to develop by this means a uniform method of presentation and style of writing. But as it is, the advantages are that everyone is involved in a learning situation where a true cross-fertilisation of ideas can be made. It is, in fact, the general editorial discussion, with specific examples of proposed teaching material in front of the meeting, that has been of the greatest value in arriving at more interesting and more relevant ways of teaching local geography. It is an inescapable process that in trying to present the material for circulation to schools an answer is sought to the question: how should local geography be taught to the older secondary non-academic pupil?

The original 'model' gave a framework for collating and writing up the material gathered during the teacher's own researches. It also tended to preserve the traditional form of textbook presentation as well as the jargon of the textbook writer. The editorial system has allowed for a process of evolution in the aims to be followed in presenting the material. Some sections of the project have been re-written several times. But at each editing the Association has come a stage nearer to establishing useful curriculum objectives for this kind of teaching.

The process is greatly aided by the trial use of the completed sections in the schools. The assessment to the extent to which the work has achieved its objectives then follows.

The Association want to produce material that stimulates the pupils into making active enquiries into the spatial relationships and landscape factors of the local environment so that when presented with information in the form of maps, diagrams, pictures, written accounts, and information derived from fieldwork excursions, he should be able to come to valid conclusions concerning these factors. The hope is to unite the methods of fieldwork with that of classwork and to rid the teaching of local geography of arid generalisations and the recitation of bald fact. Classwork becomes as much an investigation as fieldwork. The pupils' powers of observation are stimulated and directed. The material used for these investigations is the reality of the local environment and the needs of people living there today. It has been the evolutionary process of the teacher's group activity which has arrived at these conclusions, and this evolutionary process can be traced in the secretaries' files. It is not meant to imply that these are *the* curriculum objectives, but they are the objectives consistent with the need to teach the kind of pupil for whom the work is designed and also consistent with the contemporary nature of geography as a discipline.

Consequently the original model is now almost discarded and another has developed in its place. The first model was somewhat traditional in approach and followed the conventional rubric of geography teaching, laying considerable emphasis on the physical basis of the environment. This is changed now: geology is no longer chosen as the starting point for learning about the locality. This might start from the study of a pre-enclosure map, an aerial photograph of a farm, or a newspaper article describing a local industry. No longer do the teachers start by writing their account, but by assembling visual and statistical materials alongside fieldwork examples. The written content is reduced to the minimum, and the pupils' learning is through the practical investigations based on visual and fieldwork data. In short, just as the teachers have been active in preparing the material, so it is hoped the pupils will be made active in their learning.

205

Appendix: 3
A selected list of Fieldwork Centres and Bibliography

Birmingham Association of Youth Clubs,
25 Spring Road, Birmingham 15

C.H.A. Movement,
The Abbey House, Whitby, Yorkshire

Field Studies Council,
9 Devereux Court, London W.C. 2

Forestry Commission,
25 Savile Row, London W.1

The Geographical Field Group,
Bishop Grosseteste College of Education, Lincoln

Holiday Fellowship Ltd.,
142 Great North Way, London N.W.4

Methodist Guild Holidays,
2 Chester House, Pages Lane, London N.10

National Association of Youth Clubs,
30–32 Devonshire Street, London W.1

Rambler's Association,
124 Finchley Road, London N.W.3

The Scottish Field Studies Association Ltd.,
141 Bath Street, Glasgow, C.2

The Association maintains Kindrogan Field Centre, Enochdu, Blairgowrie, Perthshire. The whole district hereabouts is ideally situated for exploring nearby parts of Highland and Lowland Scotland, and particularly suited for work in geography, archaeology, local history and field biology. The Centre has good accommodation, laboratories and an expert staff. Individuals or parties of students can carry out their own special fieldwork and research and, in addition, a wide variety of courses for beginners and more advanced students are held from early March until late October.

The School Journey Association of London,
23 Southampton Place, London W.C.1

The Scottish Council of Physical Recreation,
4 Queensferry Street, Edinburgh 2

Young Men's Christian Association,
51 Victoria Street, St Albans, Herts

Young Naturalist's Association,
Red House Field Centre, Hackness, Scarborough, Yorks

Youth Hostels Association,
National Office, St Albans. Herts

Bibliography

Archer, J. E. and Dalton, T. H., *Fieldwork in Geography*, Batsford, 1968.

Bell, G. E. (Editor), *Yorkshire Field Studies*, University of Leeds, Institute of Education, 1966.

Bolton, T. and Newbury, P. A., *Geography Through Fieldwork,* Blandford, 1967.

Briault, E. W. H. and Shave, D. W., *Geography in and out of School*, Harrap, 1967.

Bull, G. B. G., *Suggestions on the Organisation of Fieldwork for 'A' Level Geography Candidates*, Associated Examining Board, 1964.

Chorley, R. J. and Haggett, P., *Frontiers in Geography Teaching*, Methuen, 1966.

Cross, M. F. and Daniel, P. A., *Fieldwork of Geography Classes*, McGraw Hill. 1968.

Dalton, R. and Thomas, S., *Lincolnshire Landscapes*, Bishop Grosseteste College, Lincoln, 1968.

Dilke, M. (Editor), *Field Studies for Schools*, Vols 1 and 2, Rivingtons, 1965.

Everson, J. *Some Aspects of Teaching Geography Through Fieldwork*, Geography Vol. 54, January, 1969.

Garnett, O., *Fundamentals in School Geography*, Harrap, 1965.

Geographical Association, *Teaching Geography in Junior Schools*, 1966.

Geographical Association Birmingham Branch, *Field Studies In The West Midlands*, 1968.

Haddon, J., *Local Geography: Geographical Survey in Rural Areas,* George Philip, 1965.

Hartop, B. B. (Editor) and others, *Durham Field Studies, A Handbook for Teachers of Geography*, University of Durham, Institute of Education, 1967.

Hoskins, W. G., *The Making of the English Landscape*, Hodder & Stoughton, 1957.

Hutchings, G. B., *Geographical Field Teaching, Geography Vol. 47*, January 1962.

Jones, P. A., *Fieldwork in Geography*, Longmans, 1969.

Long, M. (Editor), *Handbook for Geography Teachers*, Methuen, 1964.

Long, M. and Roberson, B. S., *Teaching Geography*, Heinemann, 1967.

Millward, R. and Robinson, A., *Regional Landscape Studies*, Vols 1–3, Macmillan, 1968.

Ministry of Education, *Pamphlet No. 39: Geography and Education.*

Perry, G. A., Jones, E., and Hammersley, A., *Approaches to Environmental Studies*, Blandford Press, 1968.

Rigg, J. B., *A Textbook of Environmental Science*, Constable, 1968.

Sauvain, P. A., *A Geographical Field Study Companion.* Hulton Educational Publications, 1964.

Stamp, L. D., *Applied Geography*, Penguin, 1960.

Steers, J. A. (Editor), *Field Studies in the British Isles*, Nelson, 1964.

Stimson, C. D. J. and others, *Research Publication No. 1: Local Geography in the Secondary School,* Sydney Teachers' College (Australia), 1966.

UNESCO, *Source Book for Geography Teaching*, Longmans, 1965.

University of Nottingham, Institute of Education, *Geography outside the Classroom*, 1966.

Wheeler, K. S., 'Environmental Studies' and 'Geographical Fieldwork', *Encyclopaedia of Education*, ed. Edward Blishen, Blond Educational, 1969.

Wheeler, K. S., 'Schoolchildren and the Environment'. *Town and Country Planning*, July 1966.

Wilks H. C., *Geography Fieldwork: A Continuous and Graded Course* Geography Vol 53, November 1968.

Wooldridge, S. W. *The Geographer as Scientist*, Nelson, 1956.

Yates, E. M. and Robertson, M. F. *Geographical Field Studies*, Geography, Vol. 53, January, 1968.

Fieldwork and Maps

Bridges, E. M. and Doornkamp, J. C. *Morphological Mapping and the study of Soil Patterns, Geography Vol. 48*, 1963.

Bygott, J. *Mapwork and Practical Geography*, University Tutorial Press, 1964.

Hunt, A. J. and Moisley, H. A. *Population Mapping in Urban Areas, Geography* Vol. 45, 1960

Johns, E. *The Surveying and Mapping of Vegetation and some Dartmoor Pastures, Geographical Studies IV,* 1957.

Liddell, L. E., Chapman, H. A. L. and MacFadyen, J. M.
Know the Game Series, Educational Productions, 1965.

McGregor, D. R. *'Some Observations on the Geographical Significance of Slopes',* *Geography Vol. 49*, 1964.

Sylvester, D. *Map and Landscape*, Philip, 1952.

Walters, R. S. *'Morphological Mapping, Geography Vol. 43* , 1958.

The Visual Approach to Fieldwork

Hutchings, G. E. *Introduction to Geographical Landscape Sketching*, Juniper Hall, 1958.

Hutchings, G. E. *Landscape Drawing*, Methuen, 1960.

Land Use Survey and Soil Study

Brade-Birks, S. G., *Good Soil*, Teach Yourself Series, English University Press, 1959.

Branson, J. M. *The Study of Soil*, School Nature Study Union Publication No. 22, 1950.

Buchsbaum, R. and M., *Basic Ecology*, Boxwood Press, 1967.

Clarke, G. R., *The Study of Soil in the Field*, Oxford University Press, 1957.

Coleman, A., *Land Use*, Survey Handbook, Kings College, University of London.

Hunt, A. J., *'Land-Use Survey as a Training Project',* *Geography Vol. 38*, 1953.

Mercer, I. D., Site Analysis, *Teachers World*, 24th March, 1967.

McClean, R. C. and Ivimey-Cook, W. R., *Practical Field Ecology*, Allen & Unwin, 1957.

Newbiggin, M. I., *Plant and Animal Geography*, Methuen, 1950.

Russell, Sir E. J., *Lessons on Soil*, Cambridge, 1950.

Sankey, J., *Chalkland Ecology*, Heinemann, 1966.

Sankey, J. *Guide to Field Biology*, Longmans, 1966.

Sinker, C. A., *Vegetation and the Teaching of Geography in the Field, Geography Vol. 44*, 1964.

Soil Survey of Great Britain, Field Handbook, 1960.

Taylor, J. A. *Methods of Soil Study*, Geography Vol. 45, 1960.

Tansley, A. G. *Britain's Green Mantle*, Allen & Unwin, 1949.

Waites Bryan, Wheeler, K. S., Gigg, J., *Patterns and problems In World Agriculture*, Jacaranda Press (Australia), 1970.

A Farm Study

Association of Agriculture. *Sample Farm Studies.*

Burns, A., 'Farm Survey', *Teachers World*, 1st May, 1967.

Howells, R. (Editor), *Farming in Britain*, David Rendel Ltd, 1967.

Stamp, L. D., *Man and the Land*, Collins New Naturalist Series, 1955.

Whitlock, R., *Farming from the Road*, John Baker, 1967.

Young, I. V., *Farm Studies in the Teaching of Geography*, Association of Agriculture, 1960.

Communications

Appleton, J., *The Geography of Communications in Great Britain*, O.U.P., 1961.

Bing, F. G., *The Grand Surrey Iron Railway*, Croydon Public Library Committee, 1931.

British Road Federation, *Basic Road Statistics*.

British Rail, *British Railways, Facts for Teachers*.

Buchanan Report, *Traffic in Towns* (Shortened Edition), Penguin, 1963.

Grant, W. A., *Topography of Stane Street*, John Long, 1922.

H.M.S.O., *Road Research Technical Paper 46*, D.S.I.R.

H.M.S.O., *Ministry of Transport Memo 780*.

Margary, I. D., *Roman Ways in the Weald*, Phoenix, 1949.

Ordnance Survey, *Local Accessibility*, Explanatory Text 6.

Rolt, L. T. C., *The Inland Waterways of Britain*, Allen & Unwin, 1950.

Smigielski, C., *The Leicester Traffic Plan*, Leicester Corporation, 1964.

White, H. P., *A Regional History of the Railways of Great Britain*, Vol. II: Southern England, Phoenix, 1961.

Winbolt, S. E. *With a Spade on Stane Street*, Methuen, 1932.

Fieldwork from the Air

Anonymous, Classroom on Wings, *Times Educational Supplement*, 9th September, 1966.

Beresford, M. W. and St Joseph, J. K. S. *Medieval England, An Aerial Survey*, C.U.P., 1958.

Brearley, D., *The Use of Charter Flights in the Teaching of Geography, Geography Vol. 51*, 1966.

Minshull, R., *Human Geography from the Air*, Macmillan, 1968.

Walker, F., *Geography from the Air*, Methuen, 1953.

Walton, A. D., *Teaching Geography No. 1: A Topical List of Vertical Photographs in the National Air-Photo Libraries*, Geographical Association, 1967.

Historical Geography in the Field

Atkinson, R. J. C., *Field Archaeology*, Methuen, 1963.

Barley, M. W., *English Farmhouses and Cottages*, Routledge and Kegan Paul, 1967.

Berry, B., *A Lost Roman Road*, Allen & Unwin, 1963.

Blyth, W. A. I., 'Field Studies in the Teaching of History', *Field Studies for Schools*, Vol. 1, ed. M. S. Dilke, Rivingtons, 1965.

Bowen, H. C., *Ancient Fields*, British Association, 1961.

Bowley, M., *Innovations in Building Materials*, Duckworth, 1960.

Brunskill, R. W., 'A Systematic Procedure for Recording English Vernacular Architecture', *Transactions of the Ancient Monuments Society*, Vol. 13, 1965–66.

Celoria, F., *Teach Yourself Local History*, Educational University Press.

Clifton-Taylor, Alec, *The Pattern of English Building*, Batsford, 1962.

Coppock, J. T., 'Changes in Farm and Field Boundaries in the Nineteenth Century', *Amateur Historian*, Vol. III, 1958.

Cordingley, R. A., 'British Historical Roof Types and their Members: a Classification', *Transactions of the Ancient Monuments Society*, Vol. 9, 1961.

Crawford, O. G. S., *Archaeology in the Field*, Phoenix, 1960.

Davey, N., *A History of Building Materials*, Phoenix, 1961.

Darby, H. C. (Editor), *An Historical Geography of England before 1800,* Cambridge University Press, 1963.

Domesday Geography of England, Regional Series.

Dyos, H. J., *A Victorian Suburb*, Leicester University Press, 1961.

Ekwall, E., *Concise Oxford Dictionary of English Place Names*, Oxford University Press, 1960.

Emery, F. V., 'Moated Settlements in England', *Geography Vol. 47*, 1962.

Finberg, H. P. R. and Skipp, V. H. T., *Local History*, David and Charles, 1967.

Harley, J. B., *Historians Guide to Ordnance Survey Maps*, National Council of Social Service, 1964.

Hoskins, W. G., *Leicestershire: the History of the Landscape*, Hodder & Stoughton, 1965.

Hoskins, W. G., *Fieldwork in Local History*, Faber & Faber, 1967.

Hudson, K., *Handbook for Industrial Archaeology*, John Baker, 1967.

Innocent, C. F., *The Development of English Building Construction*, 1916.

Jewell, P. A. (Editor), *The Experimental Earthwork*, British Association, 1963.

Mitchell, J. B., *Historical Geography*, English University Press, 1963.

Ordnance Survey, *Field Archaeology*, H.M.S.O., 1963.

Parnell, J. P. M., *Techniques of Industrial Archaeology*, David & Charles, 1966.

Pevsner, N., *The Buildings of England*, Regional Series, Penguin.

Sheppard, J., *Vernacular Buildings in England: a Survey of Recent Work, Trans. Institute of British Geographers No. 40*, December 1966.

Shore, B. C. G., *Stories of Britain*, Leonard Hill, 1957.

Smith, J. T., 'Medieval Roofs: a Classification', *Architects Journal 115*, 1958.

Smith, J. T., 'Cruck Construction: a survey of the problems', *Medieval Archaeology*, 8, 1964.

Steer, F. S., *Farm and Cottage Inventories*, Essex Record Office, 1950.

Thorpe, H., 'The Green Villages of County Durham', *Trans. Institute of British Geographers No. 15*, 1949.

University of Durham, Institute of Education, *History Field Studies in the Durham Area*, 1966.

Visitors Guide to Country Workshops in Britain, Rural Industries Bureau, 35 Camp Rd, Wimbledon Common, S.W.19.

Webster, G., *Practical Archaeology*, A. and C. Black, 1963.

Wheeler, K. S. and Waites, Bryan, 'Geography and Religious Education' (includes a fieldwork study of a church), *Blond's Teachers' Handbooks Series: Religious Education*, ed. Blond Educational, 1970.

Wood, E. S., *Collins' Field Guide to Archaeology* Collins, 1963.

Urban Fieldwork (Chapters 10 and 13)

Bull, G. B. G., *Fieldwork in Towns, Geography Vol. 49*, 1964.

Bull, G. B. G., *A Town Study Companion*, Hulton, 1969.

Clayton, K. M. (Editor), *Guide to London Excursions*, 20th International Congress, 1964.

Clayton, R. (Editor), *The Geography of Greater London*, Philip, 1964.

Gibbs, J. P., *Urban Research Methods*, Van Nostrans, 1965.

Hauser, P. M. (Editor), *Handbook for Social Research in Urban Areas*, UNESCO, 1964.

Hunt, A. J. and Moisley, H. A., *Population Mapping in Urban Areas, Geography Vol. 45*, 1960.

Jones, E., *Towns and Cities, Opus 13*, Oxford University Paperbacks Series, 1966.

Robertson, M. F., 'Fieldwork in Towns', *Teachers World*, 12th May, 1967.

Smailes, A. E., *The Geography of Towns*, Hutchinson University Press, 1953.

Storm, M., *Urban Growth in Britain*, The Changing World Series, Oxford University Press, 1965.

Thorpe, D. *The Geographer and Urban Studies*, Occasional Paper No. 8, Department of Geography, University of Durham, 1966.

Parish Study

Burns, A., 'A Village Study', *Teachers World*, 26th May, 1967.

Emmison, F. G., *Some Types of Common Field Parish*, National Council of Social Service, 1965.

Finberg, J., *Exploring Villages*, Routledge and Kegan Paul. 1958.

Mills, D., *The English Village*, Local Search Series, Routledge and Kegan Paul.

Stewart, C. A., *A Village Surveyed*, Arnold, 1948.

Tate, W. E., *The Parish Chest*, Cambridge University Press, 1951.

West, J., *Village Records*, Macmillan, 1962.

Fieldwork and the Young School Leaver

Cole, R., *'A Classroom Investigation of Local Industry'*, *Geography Vol. 48*, 1963.

Colledge, C. A., *'A Case Study of a Small Food Processing Factory'*, *East Midland Geographer*, December, 1966.

Geographical Association, *Geography and the Raising of the School Leaving Age*, 1966.

Moser, C. A., *Survey Methods in Social Investigation*, Heinemann Educational Books, 1967.

Newman, R. J. P., *Fieldwork Using Questionnaires and Population Data, Teaching Geography No. 6*, Geographical Association, 1969.

The Schools' Council, *Working Paper No. 11: Society and the Young School Leaver*, H.M.S.O., 1967.

Woolner, A. H., *Modern Industry in Britain*, Our Changing World Series, Oxford University Press, 1968.

The Study of a River

Dury, G. H., *The Face of the Earth*, Pelman, 1963.

Dury, G. H., *'Rivers in Geography Teaching'*, *Geography Vol. 48*, 1963.

Bunnett, R. B., *Physical Geography in Diagrams*, Longmans, 1967.

Johnson, R. H. and Paynter, J., *'The Development of a Cut-off on the River Irk at Chadderton, Lancashire'*, *Geography Vol. 52*, 1967.

Leopold, L. B. and Langbein, W. B., *'River Meanders'*, *Scientific American*, June 1966.

Sparks, B. W., *Geomorphology*, Longmans, 1960.

Wooldridge, S. W. and Linton, D. L., *Structure, Surface and Drainage in South East England*, Philip, 1955.

Coastal Fieldwork (Chapters 20 and 21)

Alexander, L. M., *Offshore Geography of North West Europe*, Murray, 1966.

Bowen, J. P., *British Lighthouses*, Longmans, 1947.

Catalogue of Admiralty Charts and Hydrographic Publications, obtainable from J. D. Potter Esq., 145 Minories, London, E.C.3.

Daysh, C. H. J. (Editor), *A Survey of Whitby and District*, Shakespeare Head, 1958.

H.M.S.O., *A Seaman's Pocket Book*, 1952.

King, C. A. M., *Beaches and Coasts*, Arnold, 1959.

MacMillan, D. H. *Tides*, C. R. Books, 1966.

Minikin, R. R., *Winds, Waves and Maritime Structure* Griffin, 1963.

Sparks, B. W., *'Geomorphology by the Seaside', Geography Vol. 39*, 1949.

Steers, J. A., *The Sea Coast*, Collin's New Naturalist Series, 1963.

Steers, J. A., *The Coastline of England and Wales,* Cambridge University Press, 1964.

Fieldwork from Kindrogan Field Centre (1 and 2).

Kay Cresswell, R., *Glaciers and Glaciation*, Hulton.

Sissons, J. B., *The Evolution of Scotland's Scenery*, Oliver & Boyd, 1967.

The Tay Valley Plan East Central (Scotland), Regional Planning Advisory Committee, 1950.

Rose, J. and McLellan, A. G., *Landforms of East Perthshire*, Scottish Field Studies Association, Annual Report 1966.

O'Dell, A. C. and Walton, K., *The Highlands and Islands of Scotland*, Nelson, 1962.

The Study of a Mountain Area

Coudry, W. M., *The Snowdonia National Park*, Collins, 1967.

Embleton, C., *Snowdonia, British Landscape Through Maps, No. 5*, Geographical Association.

Gittins, J., *'Field Studies in Snowdonia', Teachers World*, 28th April 1967.

H.M.S.O., *The Snowdonia National Park Guide.*

North, F. J., Campbell, B. and Scott, R., *Snowdonia*, New Naturalist Series, Collins, 1960.

Smith, B. and George, T., *British Regional Geology: North Wales*, H.M.S.O.

Fieldwork and Geology

Fox, C. S., *The Geology of Water Supply*, Technical Press, 1949.

Geologists' Association Guides.

Himus, G. W. and Sweeting, G. S., *The Elements of Field Geology*, University Tutorial Press, 1965.

Key Cresswell, R., *Geology for Geographers*, Hulton Education Press, 1964.

Large, N. F., *'The Pit Heap as a venue for Geographical Fieldwork'*, Geography Vol. *54*, April, 1969.

Stamp, L. D., *Britain's Structure and Scenery*, New Naturalist Series, Collins, 1961.

Trueman, A. E., *Geology and Scenery*, Penguin, 1961.

Assessing Fieldwork for the C.S.E. Examination

Archer, J. E., *'The Role of Fieldwork in the C.S.E. Examination in England and Wales'*, Geography Vol. *51*, 1966.

Edynbry, D., *Fieldwork in the Ordinary Level G.C.E. Geography*, Vol *52*, January, 1967.

H.M.S.O., *Examination Bulletin No. 1, The Certificate of Secondary Education*, 1963.

H.M.S.O., *Examination Bulletin No. 14: Trial Examinations; Geography*, 1966.

Mather, D. R., France, G. and Save, G. T., *The C.S.E., A Handbook for Moderators*, Collins, 1965.

Morris, J. A., *'The Challenge of the Certificate of Secondary Education'*, Geography Vol. *48*, 1963.

Society for Environmental Education

The aim of this Society, formed in 1968, is to provide opportunities for the discussion and exchange of ideas on the role of the environment in education. It seeks to further the development of Environmental Studies in schools as an inter-disciplinary contribution to the curriculum based on fieldwork. Conferences, courses, and meetings are arranged, and a Bulletin, edited by K. S. Wheeler, is published twice a year. Further information can be obtained from the Secretary: George Martin Esq., College of Education, Scraptoft, Leicester.

Index of Fieldwork Techniques